Physical Chemistry: Principles and Applications

Physical Chemistry: Principles and Applications

Ashok Kumar Srivastava

NY RESEARCH PRESS
New York

Published by NY Research Press
118-35 Queens Blvd., Suite 400,
Forest Hills, NY 11375, USA
www.nyresearchpress.com

Physical Chemistry: Principles and Applications
Ashok Kumar Srivastava

International Standard Book Number: 978-1-63238-623-6 (Hardback)

Cataloging-in-Publication Data

Physical chemistry : principles and applications / Ashok Kumar Srivastava.
p. cm.
Includes bibliographical references and index.
ISBN 978-1-63238-623-6
1. Chemistry, Physical and theoretical. 2. Chemistry. I. Srivastava, Ashok Kumar.
QD453.3 .P49 2019
541--dc23

Printed in China.

Contents

Preface

Physical Chemistry, as a field of study, deals with the physical properties of chemical substances. It is the study of the mechanism of chemical structures through physical concepts of energy, force, motion, etc. The theories and concepts of physical chemistry also have relevance across various fields of study such as photochemistry, material science, thermokinetics, etc. This book elucidates new techniques and their applications in a multidisciplinary approach. It attempts to understand the multiple branches that fall under the discipline. It aims to serve as a resource guide for students and experts alike and contribute to the growth of the discipline. For someone with an interest and eye for detail, this book covers the most significant topics in the field of physical chemistry.

A foreword of all chapters of the book is provided below:

Chapter 1 - Physical chemistry is a branch of chemistry that applies the principles of physics for the study of chemical systems in terms of their macroscopic, atomic or subatomic properties. The aim of this chapter is to explore the allied fields and sub-disciplines of physical chemistry, such as electrochemistry, photochemistry, equilibrium chemistry, thermochemistry, quantum chemistry, etc. which will be crucial for a holistic understanding of the subject; **Chapter 2** - Thermochemistry is a field of physical chemistry concerned with the study of heat energy in relation to chemical reactions and physical transformations. This chapter has been carefully written to provide an easy understanding of the varied aspects of chemical thermodynamics and thermochemistry, such as the fundamental principles of thermochemical equations, heat of combustion, exergonic and endergonic reactions, etc.; **Chapter 3** - Equilibrium chemistry studies systems that are in chemical equilibrium. In order to completely understand equilibrium chemistry, it is essential to understand the concepts central to it, such as chemical equilibrium, equilibrium constant and the law of chemical equilibrium, which have been discussed in detail in this chapter; **Chapter 4** - Gas is a fundamental state of matter. The properties of gases under different conditions of temperature and pressure are understood from ideal gas or real gas models. This chapter has been carefully written to provide an easy understanding of the different aspects of ideal and real gas, gas laws, kinetic theory of gases and the combined gas law; **Chapter 5** - Electrochemistry is a sub-discipline of physical chemistry that studies the inter-relationship between electricity and chemical change, with one being the cause or effect of the other. The topics elaborated in this chapter, such as oxidation and reduction, corrosion and electrolysis, will be vital for a comprehensive understanding of the subject. It also discusses in detail, the different types of electrochemical cells such as galvanic cell and electrolytic cell; **Chapter 6** - Chemical kinetics studies the rates of chemical reactions and processes. It is focused on an investigation of reaction mechanisms, transition states as well as the factors that influence reaction rates. All the diverse aspects of chemical kinetics have been carefully analyzed in this chapter, such as catalysis, reaction mechanisms and reaction rates, besides others.

At the end, I would like to thank all the people associated with this book devoting their precious time and providing their valuable contributions to this book. I would also like to express my gratitude to my fellow colleagues who encouraged me throughout the process.

Ashok Kumar Srivastava

1

An Introduction to Physical Chemistry

Physical chemistry is a branch of chemistry that applies the principles of physics for the study of chemical systems in terms of their macroscopic, atomic or subatomic properties. The aim of this chapter is to explore the allied fields and sub-disciplines of physical chemistry, such as electrochemistry, photochemistry, equilibrium chemistry, thermochemistry, quantum chemistry, etc. which will be crucial for a holistic understanding of the subject.

Physical chemistry is the branch of chemistry concerned with the interpretation of the phenomena of chemistry in terms of the underlying principles of physics. It lies at the interface of chemistry and physics, inasmuch as it draws on the principles of physics (especially quantum mechanics) to account for the phenomena of chemistry. It is also an essential component of the interpretation of the techniques of investigation and their findings, particularly because these techniques are becoming ever more sophisticated and because their full potential can be realized only by strong theoretical backing. Physical chemistry also has an essential role to play in the understanding of the complex processes and molecules characteristic of biological systems and modern materials.

That is, it includes study of the following scales of chemical properties of materials:

- Macroscopic: Macroscopic properties of substances describe how relatively large quantities of the substance behave as a group, e.g. melting points and boiling points, latent heats of fusion and vapourization, thermal conductivity, specific heat capacity, coefficient of linear thermal expansion, and many other "physical properties". Take the chemical substance H_2O as an example: A "drop of water" is liquid. At lower temperatures it freezes (becoming solid "ice") and at higher temperatures it evapourates (becoming a gas called "steam"). The physical state of H_2O as a liquid, solid, or gas defines how closely a large quantity of individual molecules of H_2O are attracted to each other. Because the molecules are very small, there is a large quantity of H_2O molecules in a single drop of water. However, these properties do not describe the chemistry of the H_2O molecule: Two hydrogen atoms and one oxygen atom form each molecule of water, regardless of its temperature, hence if it solid ice, liquid water, or gaseous steam.
- Microscopic: Microscopic properties of substances concern details of their physical properties observable only using the magnification provided by microscopes (there are, of course, different types and powers of microscopes e.g. light microscopes, electron microscopes, and more recently scanning probe

microscopes). Microscopic physical properties include, for example, the shapes and structures of crystals - which can have important consequences for the behaviour of large sections of the material of which they are a part e.g. as used in bridges, aircraft, and so on.

- Atomic: Atomic properties relate to elements. (Recall that elements consist of many individual atoms, whereas compounds consist of many molecules - which are, in turn, specific combinations of atoms joined together via chemical bonds.) Examples of atomic properties of elements include atomic numbers and atomic mass, e.g. the element Boron whose chemical symbol is B, has atomic number 5 and atomic mass 10.81 (strictly for a free neutral atom in ground state).
- Subatomic: The study of and research into subatomic "particles" (which are sometimes described in other ways, e.g. as "energy" or "waves") and their properties is often classified as part of physics, rather than chemistry. Advanced physical chemistry also includes subatomic structures and theories - which are important for nuclear chemistry.

Physical chemistry is traditionally divided into a number of disciplines, but the boundaries between them are imprecise. Thermodynamics is the study of transformations of energy. Although this study might seem remote from chemistry, in fact it is vital to the study of how chemical reactions yield work and heat. Thermodynamic techniques and analyses are also used to elucidate the tendency of physical processes (such as vaporization) and chemical reactions to reach equilibrium — the condition when there is no further net tendency to change. Thermodynamics is used to relate bulk properties of substances to each other, so that measurements of one may be used to deduce the value of another. Spectroscopy is concerned with the experimental investigation of the structures of atoms and molecules, and the identification of substances, by the observation of properties of the electromagnetic radiation absorbed, emitted, or scattered by samples. Microwave spectroscopy is used to monitor the rotations of molecules; infrared spectroscopy is used to study their vibrations; and visible and ultraviolet spectroscopy is used to study electronic transitions and to infer details of electronic structures. The enormously powerful technique of nuclear magnetic resonance is now ubiquitous in chemistry. The detailed, quantitative interpretation of molecular and solid-state structure is based in quantum theory and its use in the interpretation of the nature of the chemical bond. Diffraction studies, particularly x-ray diffraction and neutron diffraction studies, provide detailed information about the shapes of molecules, and x-ray diffraction studies are central to almost the whole of molecular biology. The scattering of neutrons, in inelastic neutron scattering, gives detailed information about the motion of molecules in liquids. The bridge between thermodynamics and structural studies is called statistical thermodynamics, in which bulk properties of substances are interpreted in terms of the properties of their constituent molecules. Another major component is chemical kinetics, the study of the rates of chemical reactions; it examines, for example, how rates of reactions respond to changes in conditions or the presence of a catalyst . Chemical kinetics is also concerned with the detailed mechanisms by which a reaction takes place, the sequences of elementary processes that convert reactants into products, including chemical reactions at solid surfaces (such as electrodes).

There are further subdivisions of these major fields. Thermochemistry is a branch of thermodynamics; its focus is the heat generated or required by chemical reactions. Electrochemistry is the study of how chemical reactions can produce electricity and how electricity can drive chemical reactions in "reverse" directions (electrolysis). Increasingly, attention is shifting from equilibrium electrochemistry (which is of crucial importance in interpreting the phenomena of inorganic chemistry) to dynamic electrochemistry, in which the rates of electron-transfer processes are the focus. Chemical kinetics has divisions that are based on the rates of reaction being studied. Special techniques for studying atomic and molecular processes on ever shorter time scales are being developed, and physical chemists are now able to explore reactions on a femtosecond (10^{-15} second) timescale. Chemical kinetics studies are theoretical as well as experimental. One goal is to understand the course of reactions in step-by-step (and atomic) detail. Techniques are available that allow investigators to study collisions between individual molecules.

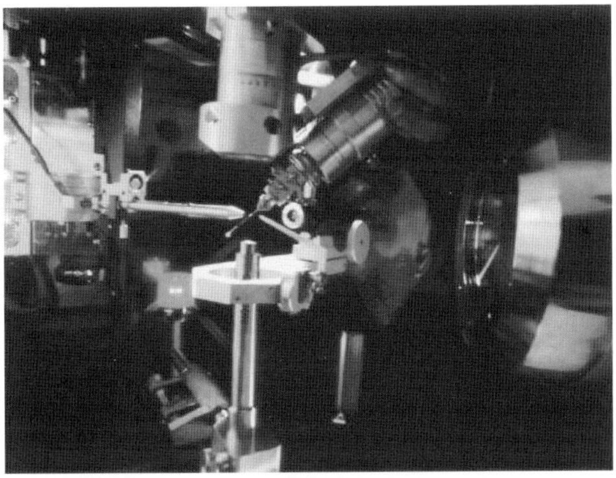

X-ray diffraction gives detailed information about shapes of molecules and is the basis of molecular biology

Physical chemistry is essential to understanding the other branches of chemistry. It provides a basis for understanding the thermodynamic influences (principally, the entropy changes accompanying reactions) that drive chemical reactions forward. It provides justifications for the schemes proposed in organic chemistry to predict and account for the reactions of organic compounds. It accounts for the structures and properties of transition metal complexes, organometallic compounds, the microporous materials known as zeolites that are so important for catalysis, and biological macromolecules, such as proteins and nucleic acids (including DNA). It is fair to say that there is no branch of chemistry (including biochemistry) that can be fully understood without interpretations provided by physical chemistry.

There is a distinction between physical chemistry and chemical physics, although the distinction is hard to define and it is not always made. In physical chemistry, the target of investigation is typically a bulk system. In chemical physics, the target is commonly an isolated, individual molecule.

Theoretical chemistry is a branch of physical chemistry in which quantum mechanics and statistical mechanics are used to calculate properties of molecules and bulk systems. The greater part of activity in quantum chemistry, as the former is commonly termed, is the computation of the electronic structures of molecules and, often, their graphical representation. This kind of study is particularly important to the screening of compounds for potential pharmacological activity, and for establishing the mode of action of enzymes.

Electrochemistry

Electrochemistry is a sub-discipline of physical chemistry; it is concerned with:

- Chemical reactions involving charged particles and electron transfers - i.e. redox reactions.
- The relationship between electrical and chemical energy and the conversion of one to the other.

Electrochemical processes are fundamental in a wide range of phenomena. For example, photosynthesis and brain chemistry are driven by electrochemical processes, as are electric batteries and fuel cells.

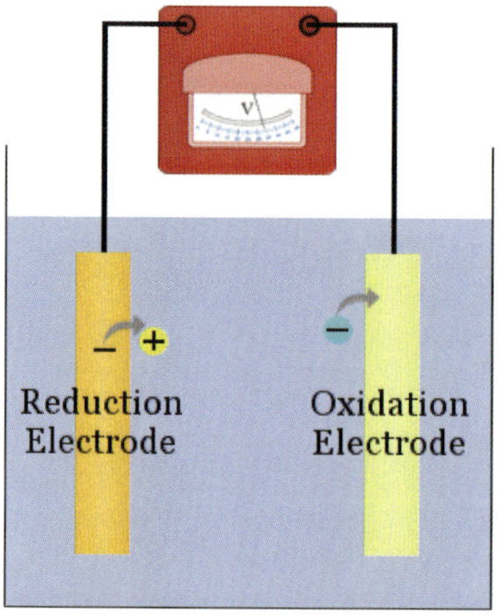

Schematic of a voltaic cell. At the left electrode an electron is transferred to a species in solution. At the right electrode a species in solution transfers an electron to the electrode

To study redox reactions, electrochemists usually find it convenient to separate the reduction and oxidation reactions in a cell using electrodes connected by a metal conductor to transport electrons from the species losing electrons to the species gaining electrons.

Equilibrium electrochemical methods allow redox potentials to be measured, from which crucial parameters such as the enthalpy, entropy, and Gibbs free energy can be determined for the reactions studied.

Chemical reaction rates can also be measured, because the reaction rate is related to the electric current - the higher the current, the faster the reaction. Techniques such as cyclic voltammetry and impedance spectroscopy may be used to determine reaction rates.

Electrochemistry is one of the main research and development disciplines involved in fields such as:

- Fuel cells
- Batteries
- Metal corrosion protection
- Electroplating

Instruments based on electrochemical methods are vital in analytical chemistry and provide convenient ways to measure, for example:

- PH
- Water contamination
- Toxic gas levels in air
- Gene expression
- Blood glucose
- Alcohol in breath

Femtochemistry

Femtochemistry is the area in the field of physical chemistry that deals with studying of the chemical reactions on a minute time scale (one femtosecond or 10^{-15} seconds). This is done to study the action of atoms within the molecules to form new molecules by rearranging themselves.

Femtochemistry is a result of the Ahmed H. Zewail's curiosity. He studied the atoms in slow motion during a chemical reaction in order to know what happens when the chemical bonds are broken to form the new bonds. Zewail even received a Nobel Prize in chemistry for his contribution in 1999.

He used an apparatus which resembled a high-speed camera. The apparatus used the laser flashes of short duration which helped in recording the time scale on which the chemical reactions takes place. The camera captured the pictures of the molecules in their transition phase when they were undergoing a chemical reaction. A femtosecond is basically 10^{-15} seconds or 0.000000000000001 seconds. Hence, this area of physical chemistry came to be known as the femtochemistry.

Importance of Femtochemistry

- It helps us in understanding the basics behind any chemical reaction. For example, femtochemistry has helped in understanding the time-resolved stages of dissociation of bromine.
- Its observations help us in understanding why certain reactions take place and why certain chemical reactions fail to complete.
- It also explains why speed and the outcome of the reaction are temperature dependent.
- Scientists use femtosecond spectrography for studying various processes in solids, liquids, gases, polymers and on the surfaces.
- The applications of femtochemistry include the functioning of catalysts, designing of the molecular electronic components, intricacy in the mechanisms of life processes along with the new innovations in the medicines.

Applications of Femtochemistry

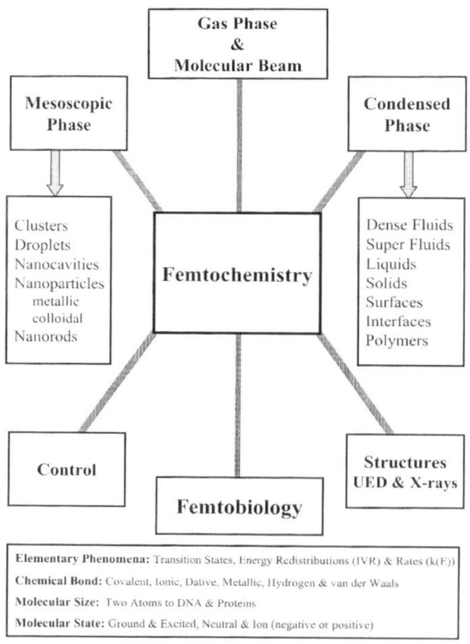

Areas of study in femtochemistry

Applications of femtochemistry have spanned the different types of chemical bonds—covalent, ionic, dative and metallic, and the weaker ones, hydrogen and van der Waals bonds. The studies have continued to address the varying complexity of molecular systems, from diatomics to proteins and DNA. Studies have also been made in the different phases of matter: gases and molecular beams; mesoscopic phases of clusters, nanostructures, particles, and droplets; condensed phases of dense fluids, liquids, solids, surfaces and interfaces; and in sibling fields of femtoscience such as femtobiology.

Photochemistry

Photochemistry is the underlying mechanism for all of photobiology. When a molecule absorbs a photon of light, its electronic structure changes, and it reacts differently with other molecules. The energy that is absorbed from light can result in photochemical changes in the absorbing molecule, or in an adjacent molecule (e.g., photosensitization). The energy can also be given off as heat, or as lower energy light, i.e., fluorescence or phosphorescence, in order to return the molecule to its ground state. Each type of molecule has a different preference for which of these different mechanisms it uses to get rid of absorbed photon energy, e.g., some prefer fluorescence over chemistry.

The Basic Laws of Photochemistry

The First Law of Photochemistry states that light must be absorbed for photochemistry to occur. This is a simple concept, but it is the basis for performing photochemical and photobiological experiments correctly. If light of a particular wavelength is not absorbed by a system, no photochemistry will occur, and no photobiological effects will be observed, no matter how long one irradiates with that wavelength of light.

The Second Law of Photochemistry states that for each photon of light absorbed by a chemical system, only one molecule is activated for a photochemical reaction. This law is true for ordinary light intensities, however, with high-powered lasers, two-photon reactions can occur, i.e., the molecule is raised to a higher energy state than produced by single-photon absorption.

The Bunsen-roscoe Law of Reciprocity states that a photochemical effect is directly proportional to the total energy dose, irrespective of the time required to deliver the dose. This law is true for chemicals in a test tube, but the response of cells to radiation usually involves a sequence of interacting biological reactions, making a linear "dose x time" relationship highly unlikely. There is no reciprocity when damage is produced, e.g., DNA damage, but there can be reciprocity over a narrow range of doses for photoreceptors that trigger a response, such as phytochrome.

Electromagnetic Radiation

Electromagnetic radiation consists of waves of electric and magnetic fields traveling in space at right angles to one another.

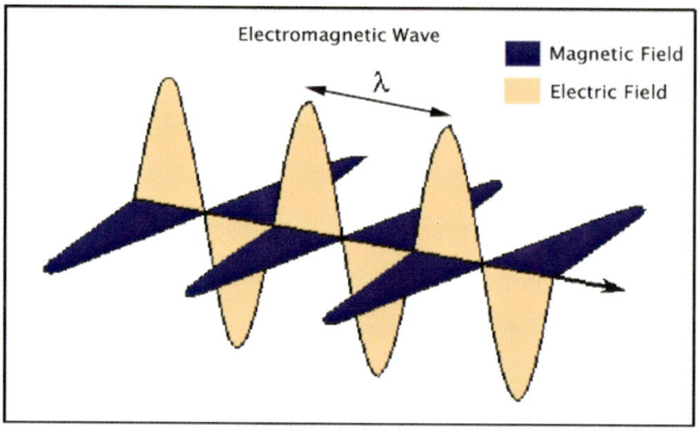

An electromagnetic wave showing the perpendicularly-oriented waves of electric and magnetic fields, and the characteristic wavelength (λ) of the radiation

The electromagnetic spectrum is composed of different wavelengths of light having different photon energies, and is classified into the regions shown in figure below. Note that the regions of interest for photochemistry, i.e., visible and ultraviolet (UV), are only a small part of the full electromagnetic spectrum. Longer wavelengths, e.g., far infrared, tend to cause the vibrational excitation of molecules, which results in heating. Shorter wavelength X-rays cause ionization.

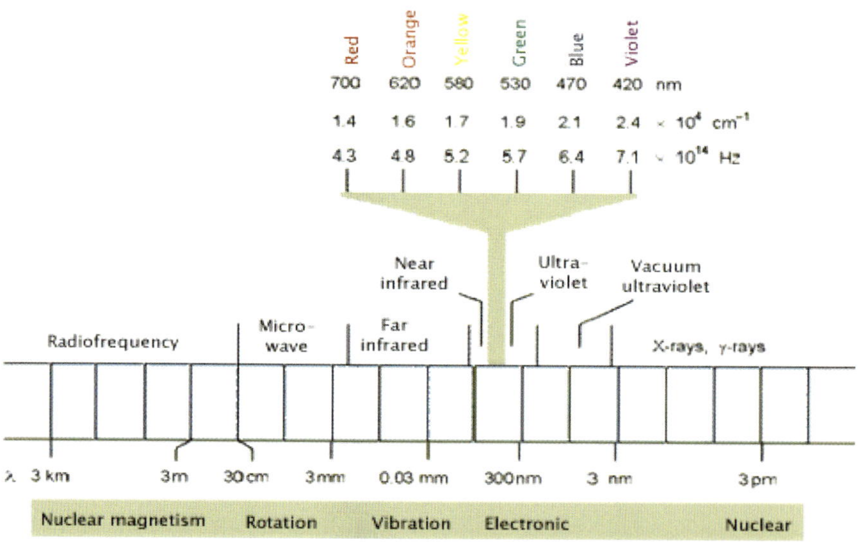

The electromagnetic spectrum highlighting the visible region, which along with the ultraviolet region, is capable of producing photochemical changes in molecules

The Jablonski Diagram

The energy gained by a molecule when it absorbs a photon causes an electron to be promoted to a higher electronic energy level. Figure illustrates the principal photophysical radiative and non-radiative processes displayed by organic molecules in solution. The symbols S_0, S_1, T_2, etc., refer to the ground electronic state (S_0), first excited singlet state (S_1), second excited triplet state (T_2), and so on. The horizontal lines represent the vibrational levels of each electronic state. Straight arrows indicate radiative transitions, and curly arrows indicate non-radiative transitions. The boxes detail the electronic spins in each orbital, with electrons shown as up and down arrows, to distinguish their spin.

Note that all transitions from one electronic state to another originate from the lowest vibrational level of the initial electronic state. For example, fluorescence occurs only from S_1, because the higher singlet states (S_2, etc.) decay so rapidly by internal conversion that fluorescence from these states cannot compete.

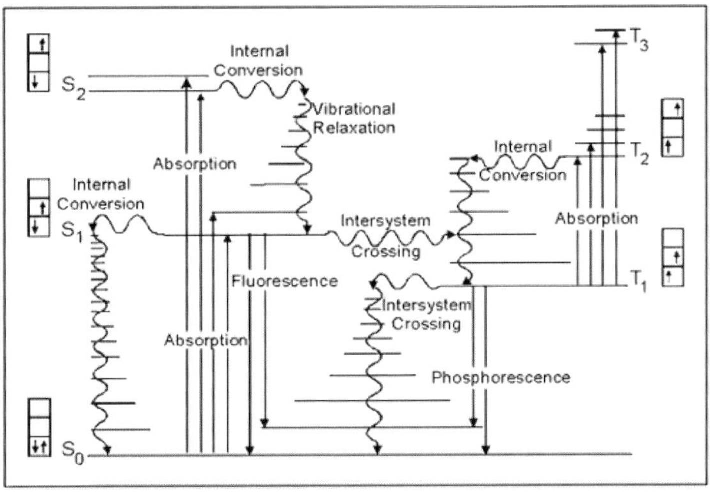

The basic concepts of this Jablonski diagram are presented in the Basic Photophysics module. This version emphasizes the spins of electrons in each of the singlet states (paired, i.e., opposite orientation, spins) compared to the triplet states (unpaired, i.e., same orientation, spins)

Electronically Excited States

The absorption of a UV or visible photon by a molecule produces an electronically excited state. The distribution of the electrons surrounding the nuclei change, as well as the forces between the atomic nuclei of a molecule. As a result, molecules in electronically excited states often have very different chemical and physical properties than their electronic ground states. For example, α -hydroxy naphthalene becomes a strong acid in its excited state. The ground state of α -hydroxy naphthalene has a pKa of 9.2, but this is reduced to 0.4 in the excited singlet state. Such molecules are known as photoacids.

The Beer-lambert Law

The absorption of photons of light is described by the Beer-lambert Law, a relationship that indicates a decrease in intensity as a beam passes through a medium that can absorb it. Consider a parallel beam of monochromatic light of initial intensity, I_o, passing through a homogeneous absorbing medium.

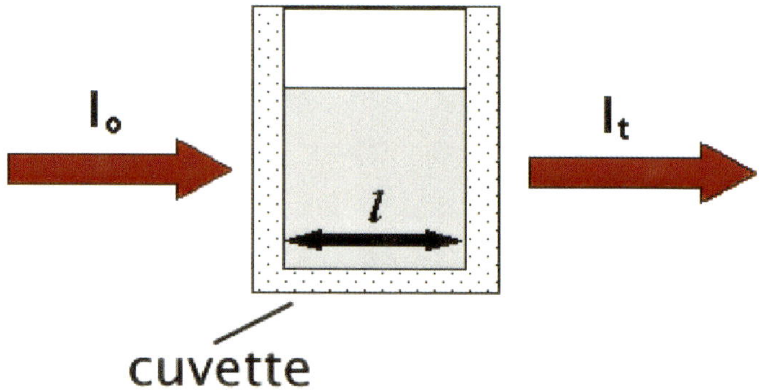

Schematic representation showing that light of initial intensity, I_o, passing through an absorbing medium in a cuvette with light path, l, will emerge with a final intensity, I_t

In spectroscopy, absorbance, A, and Optical Density, OD, are used somewhat interchangeably. Optical Density can be expressed as:

$$OD = \log_{10}(1/T)$$

where T is the transmittance. It can also be expressed as:

$$A = OD = -\log_{10}(I_t/I_o)$$

Note the log scale. An OD of 1 will have a transmittance of 0.1, and a % transmission of 10, but at OD = 2, the transmittance is 0.01, and the % transmission = 1.

Another way of expressing this information is to use the Beer-Lambert Law. It states that the absorbance, A, of a molecular species is linearly related to the path length (centimeter), l, the absorber concentration (moles/liter), c, and the proportionality constant, ε, called the molar extinction coefficient of the absorbing molecular species (liters/mole-cm) [a measure of how strongly a chemical species absorbs light at a given wavelength].

$$A = \varepsilon cl$$

Energy Level Diagram

One way to view the properties of molecular excited states is shown by the potential energy diagram in Figure. This diagram, known as a *Franck-Condon* energy level diagram, shows potential energy curves for the ground state (S_0), and first excited singlet state (S_1) of an organic molecule as a function of nuclear configuration. These curves are sometimes referred to as potential energy wells, because of their shape. The horizontal lines within each curve represent the vibrational levels of each electronic state. The lowest vibrational state for each energy level is designated as 0, and the levels above it are successively 1, 2, etc. The band assignments in brackets (*e.g.*, (0, 1)) indicate, respectively, the vibrational level of the initial state, and of the final state involved in a transition.

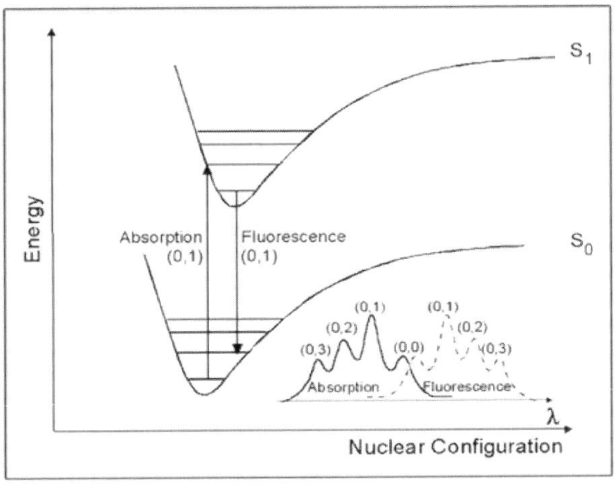

Franck-Condon Energy Level Diagram

The horizontal axis is the nuclear configuration, which can be thought of as the distance between nuclei. When considering two atoms bonded to each other, the bottom of the well corresponds to the equilibrium bond length. Because excitation involves the movement of charge density into an antibonding orbital, the equilibrium bond length in S_1 is generally longer than in S_0. This is illustrated in Figure by the displacement of the S_1 well to the right of the S_0 well.

The absorption of light takes place on a much faster time scale (~ 10^{-15} s) than molecular vibration (~ 10^{-12} s), hence the initially formed excited state must have the same nuclear configuration as the ground state. This transition is called the vertical or Franck-Condon transition, and results in the molecule having excess vibrational energy. The excess vibrational energy can be dissipated through the process of vibrational relaxation, i.e., the process of internal conversion, which returns the molecule to the lowest vibrational level of S^1. Fluorescence usually occurs from the lowest vibrational level of S^1. Because these transitions occur at lower energies than absorption, fluorescence is observed at longer wavelengths (λ) than absorption (i.e., lower energy).

Quantum Yields and Lifetimes

The energy that a molecule gains when it absorbs light is subsequently lost by a molecule in one of several ways. As shown graphically in the Jablonski diagram, it can lose the energy as heat as it returns to the ground state (internal conversion). Alternately, it can lose the energy as light (fluorescence), usually on a nanosecond time scale. A third pathway is intersystem crossing to a triplet state, from which energy can also be lost as light (phosphorescence), but over much longer times (microseconds or longer). And finally, the energy can be transferred to another molecule.

The *quantum yield* of a process is the probability that an absorbed photon undergoes one particular process. Thus, one can define a quantum yield for fluorescence, a quantum yield for phosphorescence, or a quantum yield for other pathways. Each quantum yield is typically a number between 0 and 1 (except under unusual circumstances), and the total of all quantum yields for a particular absorption event should sum to one. Note that these processes are competing. If conditions are altered such that the quantum yield for fluorescence is increased, then the quantum yield for some other process(es) must decrease.

Consider a molecule, M, that is exposed to light, and absorbs photons at the rate, I_{abs}. As shown in the following formulas, the excited singlet state of the molecule, $^1M^*$, can fluoresce emitting a photon, $h\nu f$. It can lose energy as heat, and move to the triplet excited state, $^3M^*$, by intersystem crossing (*isc*). Finally, it can lose energy as heat, and move to the ground state by internal conversion (*ic*). The rate of each of these loss processes will be proportional to the concentration of the excited singlet state, $^1M^*$, and a rate constant, *k*, for each process.

	Rate
$M \rightarrow {}^1M^*$	I_{abs}
${}^1M^* \xrightarrow{k_f} M + h\nu_f$	$k_f\left[{}^1M^*\right]$
${}^1M^* \xrightarrow{k_{isc}} {}^3M^* + \text{heat}$	$k_{isc}\left[{}^1M^*\right]$
${}^1M^* \xrightarrow{k_{ic}} M + \text{heat}$	$k_{ic}\left[{}^1M^*\right]$

Because the lifetime of the singlet excited state is relatively short, we can assume that all of the excited singlet states that are formed by light absorption will rapidly decay through one of the three means just described. This is the "steady state" approximation. Since the rate of formation of the excited singlet state is I_{abs} , then the sum of the rates of loss, each given above, must be equal to the rate of formation.

$$I_{abs} = \sum_i k_i\left[{}^1M^*\right]$$

From this relationship, one can determine the quantum yield for each process. It is simply the rate of that process as a fraction of all pathways for the loss of energy.

Intermolecular Processes: Excited State Quenching

When a second molecule (a quencher, Q) interacts with a molecule in an excited state, new ways may be created for the excited state species to lose its energy of excitation. Such interactions (collisions) can induce the loss of energy in the form of heat, which is called physical quenching, or it can cause the energy to be transferred to the second molecule with or without the transfer of an electron. The former is called energy transfer, and the latter, electron transfer. Formally, one can write:

$M^* + Q \rightarrow M + Q + heat$ physical quenching

$M^* + Q \rightarrow M + Q^*$ energy transfer

$M^* + Q \rightarrow M^{\cdot +} + Q^{\cdot -}$ electron transfer

When this kind of quenching occurs it reduces the concentration of the excited state more rapidly than if the quencher were not present. This means that fluorescence, which is proportional to the concentration of the excited state will also be reduced. By monitoring fluorescence from an excited state molecule, one can determine the concentration of a quencher, if its rate constant of interaction with the excited state is known. Once again, consider the molecule, M, that can decay by fluorescence, intersystem crossing or internal conversion. Now add a fourth possibility, quenching. The reactions and rates are then given by:

	Rate
$M \rightarrow {}^1M^*$	I_{abs}
${}^1M^* \xrightarrow{k_f} M + h\nu_f$	$k_f\left[{}^1M^*\right]$
${}^1M^* \xrightarrow{k_{isc}} {}^3M^* + heat$	$k_{isc}\left[{}^1M^*\right]$
${}^1M^* \xrightarrow{k_{ic}} M + heat$	$k_{ic}\left[{}^1M^*\right]$
${}^1M^* + Q \xrightarrow{k_q} quenching$	$k_q\left[{}^1M^*\right][Q]$

Note that the quenching reaction rate depends on 2 reactants, M and Q, and the rate constant is a bimolecular rate constant. Once again, one can make the steady state approximation, and compare the rate of fluorescence, kf[1M*] in the presence of the quencher to the same quantity in the absence of the quencher. This leads to the following relationship:

$$\frac{I_f^o}{I_f} = 1 + k_q\,\tau_f\,[Q] \qquad \text{where} \qquad I_f^o = k_f\left[{}^1M^*\right]_o \qquad I_f = k_f\left[{}^1M^*\right]$$

Where I_f^o is the intensity, or rate of fluorescence, without a quencher; I_f is the intensity, or rate of fluorescence, with a quencher; k_q is the quencher rate coefficient; τ_f is the fluorescence lifetime of A without a quencher present, and $[Q]$ is the concentration of the quencher. This relationship is referred to as the *Stern-Volmer equation.*

Types of Photochemical Reactions

1. *Linear addition to an unsaturated molecule*, e.g., the pyrimidine base, thymne, in DNA can combine with the amino acid residue, cysteine, in proteins. This is a model for the photochemical crosslinking of DNA and proteins by UV radiation.

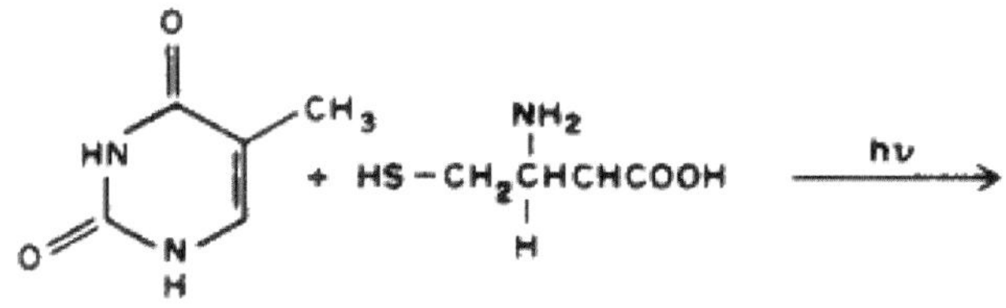

THYMINE CYSTEINE

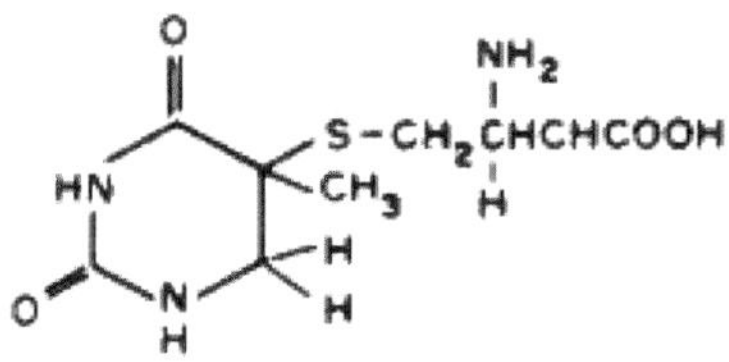

5-S-CYSTEINYL-6-HYDROTHYMINE

2. *Cycloaddition of unsaturated molecules*, e.g., two thymines can react to form a ring product, the thymine dimer, an important class of products formed in DNA by UV radiation.

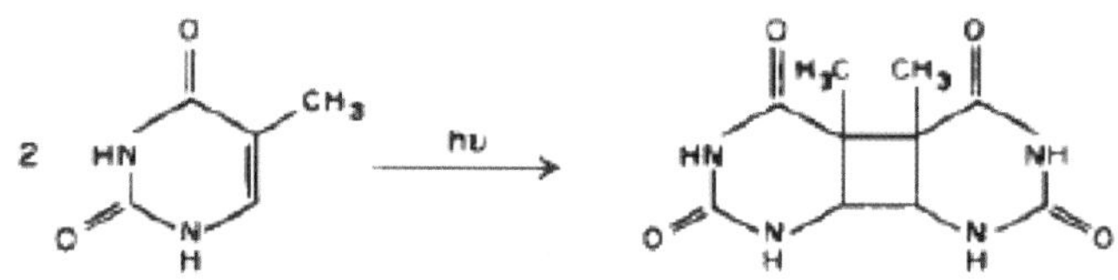

THYMINE THYMINE DIMER

3. *Photofragmentation*, e.g., the side chain of riboflavin can split off to form lumiflavin.

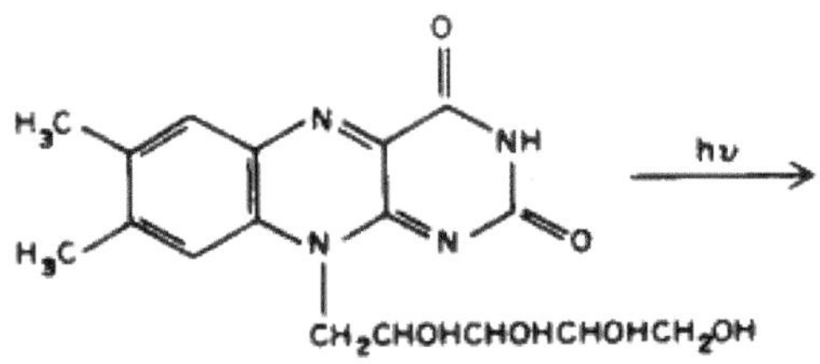

RIBOFLAVIN

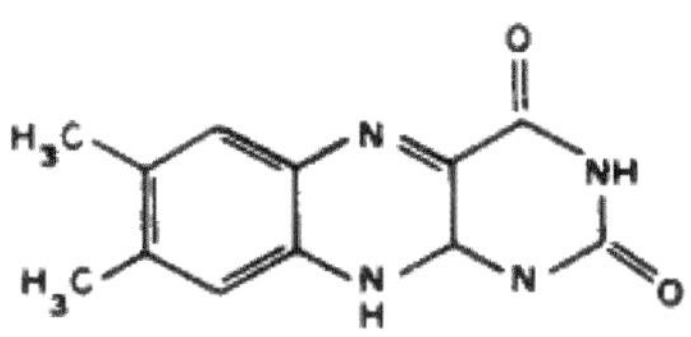

LUMIFLAVIN

+ $CH_3\overset{O}{\overset{\|}{C}}CHOHCHOHCH_2OH$

4. *Photooxidation*, e.g., the ring structure of cholesterol can add a peroxy group.

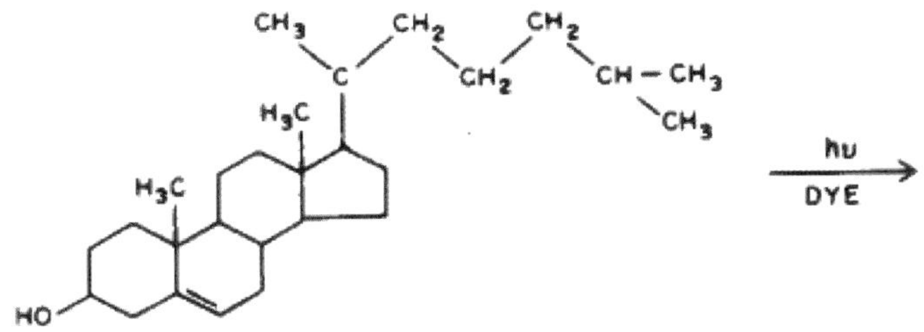

CHOLESTEROL

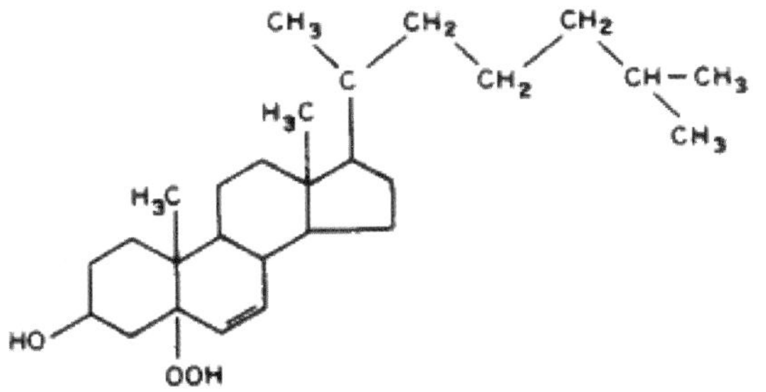

3β-HYDROXY-5α-HYDROPEROXY-Δ⁶-CHOLESTENE

5. *Photohydration*, e.g., uracil can add a molecule of water to it 5-6 double bond when UV irradiated.

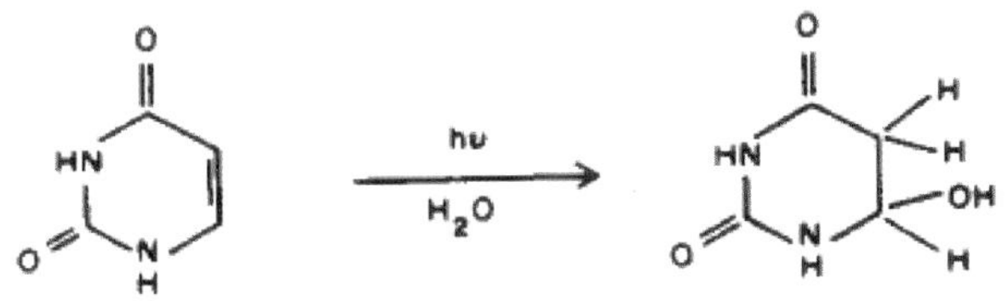

URACIL 6-HYDROXY-5-HYDROURACIL

6. *Cis-Trans Isomerization*, e.g., all-*trans* retinal can be converted to 11-*cis* retinal.

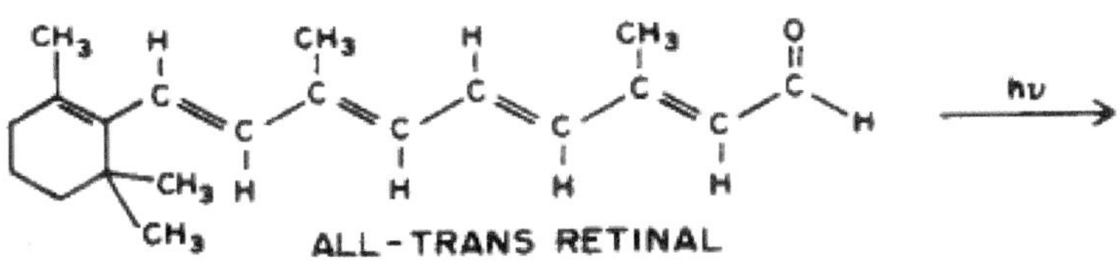

ALL-TRANS RETINAL

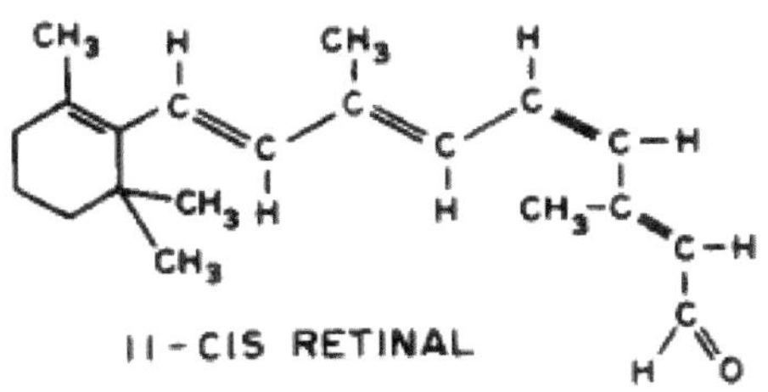

11-CIS RETINAL

7. *Photorearrangement*, e.g., 7-dehydrocholesterol can be converted to vitamin D_3.

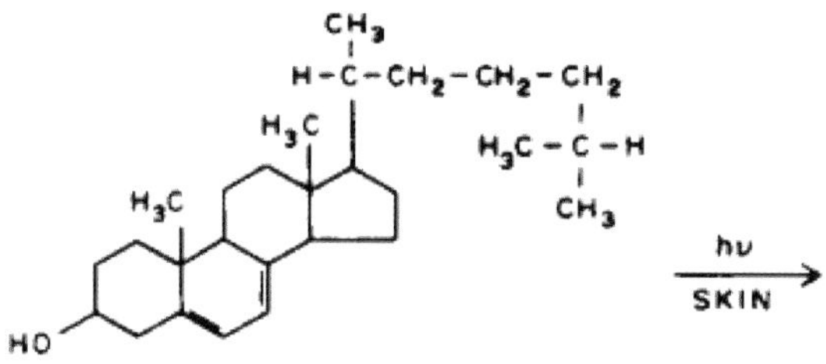

7-DEHYDROCHOLESTEROL

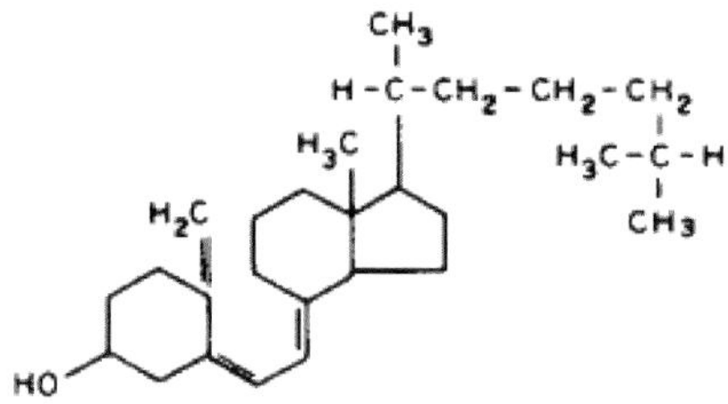

VITAMIN D_3

8. *Energy Transfer*, e.g., all photosensitized reactions

Organic Photochemistry

Examples of photochemical organic reactions are electrocyclic reactions, radical reactions, photoisomerization and Norrish reactions.

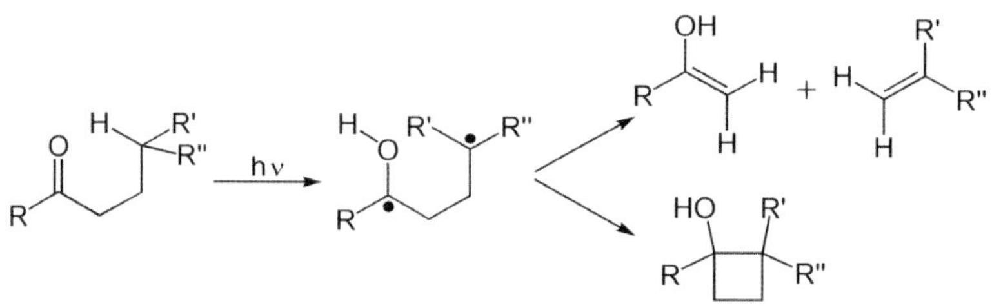

Alkenes undergo many important reactions that proceed via a photon-induced π to π* transition. The first electronic excited state of an alkene lack the π-bond, so that rotation about the C-C bond is rapid and the molecule engages in reactions not observed thermally. These reactions include cis-trans isomerization, cycloaddition to other (ground state) alkene to give cyclobutane derivatives. The cis-trans isomerization of a (poly)alkene is involved in retinal, a component of the machinery of vision. The dimerization of alkenes is relevant to the photodamage of DNA, where thymine dimers are observed upon illuminating DNA to UV radiation. Such dimers interfere with transcription. The beneficial effects of sunlight are associated with the photochemically induced retro-cyclization (decyclization) reaction of ergosterol to give vitamin D. In the DeMayo reaction, an alkene reacts with a 1,3-diketone reacts via its enol to yield a 1,5-diketone. Still another common photochemical reaction is Zimmerman's Di-pi-methane rearrangement.

In an industrial application, about 100,000 tonnes of benzyl chloride are prepared annually by the gas-phase photochemical reaction of toluene with chlorine. The light is absorbed by chlorine molecule, the low energy of this transition being indicated by the yellowish color of the gas. The photon induces homolysis of the Cl-Cl bond, and the resulting chlorine radical converts toluene to the benzyl radical:

$$Cl_2 + h\nu \rightarrow 2\ Cl\cdot$$

$$C_6H_5CH_3 + Cl\cdot \rightarrow C_6H_5CH_2\cdot + HCl$$

$$C_6H_5CH_2\cdot + Cl\cdot \rightarrow C_6H_5CH_2Cl$$

Mercaptans can be produced by photochemical addition of hydrogen sulfide (H_2S) to alpha olefins.

Inorganic and Organometallic Photochemistry

Coordination complexes and organometallic compounds are also photoreactive. These reactions can entail cis-trans isomerization. More commonly photoreactions result in dissociation of ligands, since the photon excites an electron on the metal to an orbital that is antibonding with respect to the ligands. Thus, metal carbonyls that resist thermal substitution undergo decarbonylation upon irradiation with UV light. UV-irradiation of a THF solution of molybdenum hexacarbonyl gives the THF complex, which is synthetically useful:

$$Mo(CO)_6 + THF \rightarrow Mo(CO)_5(THF) + CO$$

In a related reaction, photolysis of iron pentacarbonyl affords diiron nonacarbonyl:

$$2\ Fe(CO)_5 \rightarrow Fe_2(CO)_9 + CO$$

Select photoreactive coordination complexes can undergo oxidation-reduction processes via single electron transfer. This electron transfer can occur within the inner or outer coordination sphere of the metal.

Examples of Photochemical Reactions

- Photosynthesis: plants use solar energy to convert carbon dioxide and water into glucose and oxygen.
- Human formation of vitamin D by exposure to sunlight.
- Bioluminescence: *e.g.* In fireflies, an enzyme in the abdomen catalyzes a reaction that produced light.
- Polymerizations started by photoinitiators, which decompose upon absorbing light to produce the free radicals for radical polymerization.

- Photodegradation of many substances, e.g. polyvinyl chloride and Fp. Medicine bottles are often made with darkened glass to prevent the drugs from photodegradation.
- Photodynamic therapy: light is used to destroy tumors by the action of singlet oxygen generated by photosensitized reactions of triplet oxygen. Typical photosensitizers include tetraphenylporphyrin and methylene blue. The resulting singlet oxygen is an aggressive oxidant, capable of converting C-H bonds into C-OH groups.
- Photoresist technology, used in the production of microelectronic components.
- Vision is initiated by a photochemical reaction of rhodopsin.
- Toray photochemical production of ε-caprolactame.
- Photochemical production of artemisinin, anti-malaria drug.
- Photoalkylation, used for the light-induced addition of alkyl groups to molecules.

Equilibrium Chemistry

Equilibrium chemistry is concerned with systems in chemical equilibrium. The unifying principle is that the free energy of a system at equilibrium is the minimum possible, so that the slope of the free energy with respect to the reaction coordinate is zero. This principle, applied to mixtures at equilibrium provides a definition of an equilibrium constant. Applications include acid–base, host–guest, metal–complex, solubility, partition, chromatography and redox equilibria.

Thermodynamic Equilibrium

Thermodynamic equilibrium, condition or state of a thermodynamic system, the properties of which do not change with time and that can be changed to another condition only at the expense of effects on other systems. For a thermodynamic equilibrium system with given energy, the entropy is greater than that of any other state with the same energy. For a thermodynamic equilibrium state with given pressure and temperature, the Gibbs free energy is smaller than that of any other state with the same pressure and temperature.

Equilibrium Constant

The equilibrium constant is the ratio of the equilibrium concentrations of the products raised to the power of their stoichiometric coefficients to the equilibrium concentrations of the reactants raised to the power of their stoichiometric coefficients.

For a reversible reaction:

$$aA + bB \rightarrow cC + dD$$

The equilibrium constant, K, is equal to:

$$K = [C]^c \cdot [D]^d / [A]^a \cdot [B]^b$$

where

[A] = equilibrium concentration of A

[B] = equilibrium concentration of B

[C] = equilibrium concentration of C

[D] = equilibrium concentration of D

Thermochemistry

Thermochemistry is the study of heat and energy associated with a chemical reaction or a physical transformation. A reaction may absorb or release energy. The reaction which absorbs energy in the form of heat is called endothermic reaction and that which releases energy in the form of heat is called exothermic reaction.

Measurement of heat changes are carried out in vessels called calorimeters. Metallic calorimeters are not used for measuring thermochemical changes because metals may react with substances. Stainless steel or gold plated copper calorimeters may be used. Reactions may also be carried out in beakers placed in a thermo flask, in a thermally insulated box or in a styrofoam cup. During measurement of heat changes, the calorimeter also absorbs some heat; this amount of heat should also be known. It is called calorimeter constant or water equivalent of calorimeter. To determine the water equivalent of calorimeter, known volume of hot water at a specified temperature is added to known volume of water contained in the calorimeter at room temperature. Since energy is conserved, the heat taken by calorimeter and cold water should be equal to heat given by hot water. Thus, we can write the following equation,

Heat given out by hot water = Heat taken by calorimeter and cold water

$$m_1(t_2 - t_3) = (m_2 + W)(t_3 - t_1)$$

$$(m_2 + W) = m_1 \frac{(t_2 - t_3)}{(t_3 - t_1)}$$

Water equivalent of calorimeter, $W = m_1 \frac{(t_2 - t_3)}{(t_3 - t_1)} - m_2$

m_1 = *Mass of hot water*

m_2 = *Mass of cold water*

t_1 = *Temperature of cold water*

t_2 = *Temperature of hot water*

t_3 = *Temperature after mixing*

It is known that when a solute is dissolved in a solvent, heat is absorbed or evolved. Thus, dissolution of a solute in a solvent is accompanied by enthalpy change (ΔH) of the system. If heat is absorbed, ΔH is positive. If heat is evolved, ΔH is negative.

The enthalpy change per mole of a solute dissolved varies with the concentration of the solution. The enthalpy of dissolution is defined as the enthalpy change per mole of a solute when it is dissolved in a pure solvent to give a solution of specified concentration.

For example, when one mole of anhydrous calcium chloride is dissolved in 400 moles of water, 78.60 kJ heat is evolved. The thermochemical equation is written as:

$$CaCl_2\ (s) + 400H_2O(l) \rightarrow CaCl_2\ (400H_2O)\ \Delta H = -78.60\ kJ$$

Similarly, when one mole of hydrated calcium chloride is dissolved in 400 moles of water, 18.83 kJ heat is absorbed. The thermochemical equation is written as:

$$CaCl_2.6H_2O(s) + 400H_2O(l) \rightarrow CaCl_2.6H_2O(400H_2O)\Delta H = +18.83kJ$$

Solid-state Chemistry

Solid-state chemistry, also sometimes referred to as materials chemistry, is the study of the synthesis, structure, and properties of solid phase materials, particularly, but not necessarily exclusively of, non-molecular solids. It therefore has a strong overlap with solid-state physics, mineralogy, crystallography, ceramics, metallurgy, thermodynamics, materials science and electronics with a focus on the synthesis of novel materials and their characterisation.

Synthetic Methods

Given the diversity of solid state compounds, an equally diverse array of methods are used for their preparation. For organic materials, such as charge transfer salts, the methods operates near room temperature and are often similar to the techniques of organic synthesis. Redox reactions are sometimes conducted by electrocrystallisation.

Oven Techniques

For thermally robust materials, high temperature methods are often employed. For example, bulk solids are prepared using tube furnaces, which allow reactions to be conducted up to ca. 1100°C. Special equipment e.g. ovens consisting of a tantalum tube through which an electric current is passed can be used for even higher temperatures up to 2000°C. Such high temperatures are at times required to induce diffusion of the reactants, but this depends strongly on the system studied. Some solid state reactions already proceed at temperatures as low as 100°C.

Melt Methods

One method often employed is to melt the reactants together and then later anneal the solidified melt. If volatile reactants are involved the reactants are often put in an ampoule that is evacuated-often while setting the reactant mixture cold e.g. by keeping the bottom of the ampoule in liquid nitrogen-and then sealed. The sealed ampoule is then put in an oven and given a certain heat treatment.

Solution Methods

It is possible to use solvents to prepare solids by precipitation or by evaporation. At times the solvent is used hydrothermally, i.e. under pressure at temperatures higher than the normal boiling point. A variation on this theme is the use of flux methods, where a salt of relatively low melting point is added to the mixture to act as a high temperature solvent in which the desired reaction can take place.

Gas Reactions

Many solids react vigorously with reactive gas species like chlorine, iodine, oxygen etc. Others form adducts with other gases, e.g. CO or ethylene. Such reactions are often carried out in a tube that is open ended on both sides and through which the gas is passed. A variation of this is to let the reaction take place inside a measuring device such as a TGA. In that case stoichiometric information can be obtained during the reaction, which helps identify the products.

A special case of a gas reaction is a chemical transport reaction. These are often carried out in a sealed ampoule to which a small amount of a transport agent, e.g. iodine is added. The ampoule is then placed in a zone oven. This is essentially two tube ovens attached to each other which allows a temperature gradient to be imposed. Such a method can be used to obtain the product in the form of single crystals suitable for structure determination by X-ray diffraction.

Chemical vapour deposition is a high temperature method that is widely employed for the preparation of coatings and semiconductors from molecular precursors.

Air and Moisture Sensitive Materials

Many solids are hygroscopic and/or oxygen sensitive. Many halides e.g. are very 'thirsty' and can only be studied in their anhydrous form if they are handled in a glove box filled with dry (and/or oxygen-free) gas, usually nitrogen.

Characterization

New Phases, Phase Diagrams, Structures

The synthetic methodology and the characterization of the product often go hand in hand in the sense that not one but a series of reaction mixtures are prepared and subjected to heat treatment. The stoichiometry is typically *varied* in a systematic way to find which stoichiometries will lead to new solid compounds or to solid solutions between known ones. A prime method to characterize the reaction products is powder diffraction, because many solid state reactions will produce polycristalline ingots or powders. Powder diffraction will facilitate the identification of known phases in the mixture. If a pattern is found that is not known in the diffraction data libraries an attempt can be made to index the pattern, i.e. to identify the symmetry and the size of the unit cell. (If the product is not crystalline the characterization is typically much more difficult.)

Once the unit cell of a new phase is known, the next step is to establish the stoichiometry of the phase. This can be done in a number of ways. Sometimes the composition of the original mixture will give a clue, if one finds only one product - a single powder pattern - or if one was trying to make a phase of a certain composition by analogy to known materials but this is rare. Often considerable effort in refining the synthetic methodology is required to obtain a pure sample of the new material. If it is possible to separate the product from the rest of the reaction mixture elemental analysis can be used. Another way involves SEM and the generation of characteristic X-rays in the electron beam. The easiest way to solve the structure is by using single crystal X-ray diffraction.

The latter often requires *revisiting* and refining the preparative procedures and that is linked to the question which phases are stable at what composition and what stoichiometry. In other words, what does the phase diagram looks like. An important tool in establishing this is thermal analysis techniques like DSC or DTA and increasingly also, thanks to the advent of synchrotrons temperature-dependent powder diffraction. Increased knowledge of the phase relations often leads to further refinement in synthetic procedures in an iterative way. New phases are thus characterized by their melting points and their stoichiometric domains. The latter is important for the many solids that are non-stoichiometric compounds. The cell parameters obtained from XRD are particularly helpful to characterize the homogeneity ranges of the latter.

Further Characterization

In many (but certainly not all) cases new solid compounds are further characterized by a variety of techniques that straddle the fine line that (hardly) separates solid-state chemistry from solid-state physics.

Optical Properties

For non-metallic materials, it is often possible to obtain UV/VIS spectra. In the case of semiconductors that will give an idea of the band gap.

Quantum Chemistry

Quantum chemistry is the application of quantum mechanical principles and equations to the study of molecules. In order to understand matter at its most fundamental level, we must use quantum mechanical models and methods. There are two aspects of quantum mechanics that make it different from previous models of matter. The first is the concept of wave-particle duality; that is, the notion that we need to think of very small objects (such as electrons) as having characteristics of both particles and waves. Second, quantum mechanical models correctly predict that the energy of atoms and molecules is always quantized, meaning that they may have only specific amounts of energy. Quantum chemical theories allow us to explain the structure of the periodic table, and quantum chemical calculations allow us to accurately predict the structures of molecules and the spectroscopic behavior of atoms and molecules.

Wave-particle Duality for Light

Quantum mechanical ideas began with studies of the physics of light. By the late nineteenth century, virtually all scientists believed that light behaved as a wave. Although some earlier scientists, such as Isaac Newton in the seventeenth century, had thought of light as consisting of particles, the early nineteenth-century experiments of Thomas Young and Augustin Fresnel demonstrated that light has wavelike properties. In these experiments, light was passed through a pair of slits in a screen, and produced alternating light and dark regions (interference patterns) on a second screen. This phenomenon, known as diffraction, cannot be explained using a particle model for light. In the late nineteenth century, James Clerk Maxwell derived a set of equations based on the wave model for light, which beautifully explained most experimental results.

Despite this apparent certainty that light was a wave, Max Planck and Albert Einstein, at the beginning of the twentieth century, showed that some experiments required the use of a particle model for light, rather than a wave model. Since both models were necessary for an accurate description of all of the properties of light, scientists today use mathematical equations appropriate to both waves and particles in describing the properties of light.

Waves and particles are fundamentally different, a particle exists at a particular point in space, whereas a wave continues on for (sometimes) a great distance. It defies intuition to think that both of these models might describe the same thing. Nevertheless, an accurate description of light requires the use of both wave and particle ideas.

The Wave Nature of Matter

The success of wave-particle duality in describing the properties of light paved the way for using that same idea in describing matter. Experiments in the early twentieth century showed that the energy in atoms is quantized*; that is, a given atom can have only specific amounts of energy. For hydrogen, the simplest of the atoms, an accurate formula for the possible energies had been experimentally determined but was unexplainable using any particle model for the atom. The best picture that the particle model could give, consistent with experiments on atoms, put the electron in a sort of "orbit" around the nucleus. Unfortunately, the particle model predicts that the electron should collide with the nucleus, releasing energy in the process. Obviously there was a need for a different model for the electron.

* One example of quantization would be to think of your distance from the ground when standing on a ladder—your distance from the ground can only change one rung at a time.

In 1924 Louis de Broglie presented a theory for the hydrogen atom that modeled the electron as a wave. Calculations made for this model give the quantization of energy that is experimentally observed in this atom. De Broglie also postulated a general formula for obtaining the wavelength of a moving object. His formula, which is analogous to that used for light, states that the wavelength of a moving object is inversely proportional to its momentum (mass times velocity). When one uses de Broglie's formula to determine wavelengths of macroscopic objects, one discovers that the wavelengths of even the smallest objects visible to the naked eye are too small for the wavelike characteristics of these objects to be significant in any real situation. For the electron, however, the wavelength is large enough to be measurable. Diffraction experiments have been performed using electrons, demonstrating conclusively that they have wave properties.

It is contrary to our intuition that electrons might behave as waves. The repercussions of this notion are that the electron does not have a definite size, but is spread out over a region in space. We are more comfortable with the thought of the electron being a microscopic particle, moving around in an orbit near the nucleus of an atom. As with light, however, we do not abandon the particle model for electrons; rather, we employ mathematical equations arising from both particle and wave models. For quantum chemical calculations, the wave model turns out to be more useful.

The Heisenberg Uncertainty Principle

One consequence of the wave nature of matter is that the position and momentum of small objects are not well known, as they would be for a particle model. In some circumstances, a wave may be confined to a very narrow region in space; however, there is still some uncertainty as to its position. Additionally, the value of the momentum of a quantum object is often not known precisely. In 1927 Werner Heisenberg showed that the product of the uncertainty in position and the uncertainty in momentum is greater than or equal to a certain constant (Planck's constant divided by 4π). This constant is very small; accordingly, quantum mechanical uncertainty in position and momentum of objects that are large enough to see is not noticed experimentally. For electrons, however, quantum mechanical uncertainties in position and momentum are important considerations in interpreting both theoretical models and experimental results. The relationship between the uncertainties in position and momentum is known as the Heisenberg Uncertainty Principle. It tells us that the more we know about the position of a small object, such as an electron, the less we know about its momentum (and vice versa).

Calculating the Wavefunction

For a scientist, knowing that matter behaves as a wave is useful only if one knows something about that wave. The wavefunction is a mathematical function describing the wave. For example, $y(x) = A \sin(kx)$ might be the wave-function for a one-dimensional wave, which exists along the x-axis. Matter waves are three-dimensional;

the relevant wavefunction depends on the *x, y,* and *z* coordinates of the system being studied (and sometimes on time as well). We conventionally label the wavefunction for a three-dimensional object as $\psi(x, y, z)$. In 1926 Erwin Schrödinger introduced a mathematical equation whereby, if one knows the potential energy acting on an object, one can calculate the wavefunction for that object. Heisenberg had already introduced a mathematical formalism for performing quantum mechanics calculations, without explicitly including the concept of waves. It was later shown that, although the approaches of Schrödinger and Heisenberg looked very different, they made exactly the same predictions. In practice, the Schrödinger formalism is more useful for explaining the problem being studied, and the Heisenberg methodology allows for more straightforward computation. Accordingly, a mixture of the two approaches is typically used in modern quantum chemistry. Once we know the wavefunction of the atom or molecule under study, we can calculate the properties of that atom or molecule.

Quantum Mechanics of Atoms

An exact solution for Schrödinger's wave equation can be obtained for the hydrogen atom; however, for larger atoms and molecules (which contain more than one electron), Schrödinger's equation can be solved only approximately. Although this may sound so restrictive as to make the equation useless, there are well-established approaches that allow for practical and accurate calculations on atoms and molecules. This is done by making some assumptions about larger systems based upon the hydrogen atom.

When the Schrödinger equation is solved for the hydrogen atom, the resulting wavefunctions for the various possible energies that the atom can have are used to determine atomic orbitals. An orbital is a region in space where an electron is most likely to be found. For example, the lowest-energy wavefunction for a hydrogen atom is the so-called 1s orbital, which is a spherical region in space surrounding the nucleus. For some higher-energy states, the orbitals are not necessarily spherical in shape.

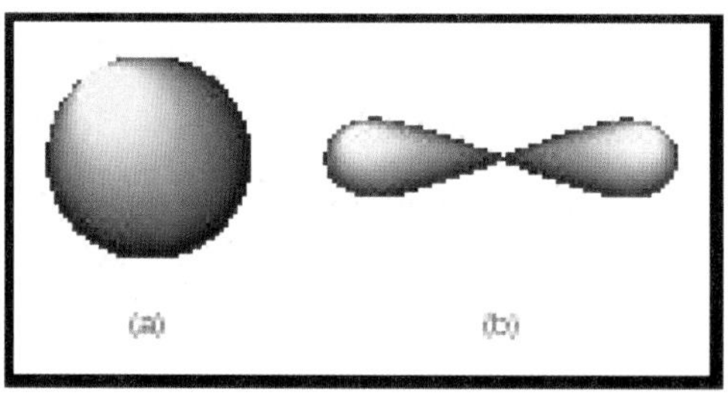

For atoms larger than hydrogen, one assumes that the orbitals occupied by the electrons have the same shape as the hydrogen orbitals but are differing in size and energy. The energies corresponding to these orbitals may be found by solving an approximate version of Schrödinger's equation. These atomic orbitals, in turn, may be used as the building blocks to the electronic behavior in molecules. As it happens, two electrons may share an atomic orbital; we say that these electrons are paired. Chemists have developed a system of rules for determining which orbitals are occupied in which atoms; calculations can then be done to determine the energies of the electrons in the atoms.

For atoms larger than hydrogen, one assumes that the orbitals occupied by the electrons have the same shape as the hydrogen orbitals but are differing in size and energy. The energies corresponding to these orbitals may be found by solving an approximate version of Schrödinger's equation. These atomic orbitals, in turn, may be used as the building blocks to the electronic behavior in molecules. As it happens, two electrons may share an atomic orbital; we say that these electrons are paired. Chemists have developed a system of rules for determining which orbitals are occupied in which atoms; calculations can then be done to determine the energies of the electrons in the atoms.

Quantum Mechanics of Molecules

Molecules are held together by covalent bonds. The simplest definition of a covalent bond is a shared pair of electrons. There are two basic approaches to modeling covalent bonds in molecules: the valence bond model and the molecular orbital model. In the valence bond model, we think of atomic orbitals on each of two atoms combining to form a bond orbital, with one electron from each atom occupying this orbital. Both the bond orbital and the electron pair now "belong" to both of the atoms. This sharing of electrons brings about a lowering in the energy, which makes the formation of molecules from atoms an energetically favorable process. The valence bond model gives the simplest quantum mechanical picture of chemical bonding, but it is not the best method for accurate calculations on molecules containing more than two atoms.

Molecular orbital theory differs from valence bond theory in that it does not require the electrons involved in a bond to be localized between two of the atoms in a molecule. Instead, the electron occupies a molecular orbital, which may be spread out over the entire molecule. As in the valence bond approach, the molecular orbital is formed by adding up contributions from the atomic orbitals on the atoms that make up the molecule. This approach, which does not explicitly model bonds as existing between two atoms, is somewhat less appealing to the intuition than the valence bond approach. However, molecular orbital calculations typically yield better predictions of molecular structure and properties than valence bond methods. Accordingly, most commercially available quantum chemistry software packages rely on molecular orbital methods to perform calculations.

Schrödinger's Wave Equation

Schrödinger's Wave equation may be written (in abbreviated form) as:

$$\hat{E}_K \psi(x, y, z) + \hat{E}_P \psi(x, y, z) = \hat{E} \psi(x, y, z)$$

The first term, $\hat{E}_K \psi (x, y, z)$, represents the kinetic energy of the system being studied. The second term, $\hat{E}_P \psi (x, y, z)$, represents the potential energy of the system. E and $\psi(x, y, z)$ are the total energy of the system and wavefunction describing the system, respectively. Once the wavefunction is determined, virtually any property of the molecule may be calculated.

A lot of the modern research in quantum chemistry is focused on improving the valence bond and molecular orbital methods for calculating molecular properties. Different underlying approximations and different orbital functions are tried, and the results are compared with previous calculations and with experimental data to determine which methods give the best results. It is often the case that the best choice of quantum chemical method depends on the particular molecule or molecular property being studied.

Chemical Dynamics

A further step can consist of solving the Schrödinger equation with the total molecular Hamiltonian in order to study the motion of molecules. Direct solution of the Schrödinger equation is called *quantum molecular dynamics*, within the semiclassical approximation *semiclassical molecular dynamics*, and within the classical mechanics framework *molecular dynamics (MD)*. Statistical approaches, using for example Monte Carlo methods, are also possible.

Adiabatic Chemical Dynamics

In adiabatic dynamics, interatomic interactions are represented by single scalar potentials called potential energy surfaces. This is the Born–Oppenheimer approximation introduced by Born and Oppenheimer in 1927. Pioneering applications of this in chemistry were performed by Rice and Ramsperger in 1927 and Kassel in 1928, and generalized into the RRKM theory in 1952 by Marcus who took the transition state theory developed by Eyring in 1935 into account. These methods enable simple estimates of unimolecular reaction rates from a few characteristics of the potential surface.

Non-adiabatic Chemical Dynamics

Non-adiabatic dynamics consists of taking the interaction between several coupled potential energy surface (corresponding to different electronic quantum states of the molecule). The coupling terms are called vibronic couplings. The pioneering work in this field was done by Stueckelberg, Landau, and Zener in the 1930s, in their work on what is now known as the Landau–Zener transition. Their formula allows the transition probability between two diabatic potential curves in the neighborhood of an avoided crossing to be calculated. Spin-forbidden reactions are one type of non-adiabatic reactions where at least one change in spin state occurs when progressing from reactant to product.

Surface Chemistry

Surface science is the study of chemical phenomena that occur at the interface of two phases, including solid-liquid interfaces, solid–gas interfaces, solid–vacuum interfaces, and liquid-gas interfaces. It includes the fields of surface chemistry and surface physics. Some related practical applications are classed as surface engineering. The science encompasses concepts such as heterogeneous catalysis, semiconductor device fabrication, fuel cells, self-assembled monolayers, and adhesives. Surface science is closely related to interface and colloid science. Interfacial chemistry and physics are common subjects for both. The methods are different. Surface chemistry can also be roughly defined as the study of chemical reactions at interfaces. It is closely related to surface engineering, which aims at modifying the chemical composition of a surface by incorporation of selected elements or functional groups that produce various desired effects or improvements in the properties of the surface or interface.

Surface chemistry is important in many critical chemical processes, such as enzymatic reactions at biological interfaces found in cell walls and membranes, in electronics at the surfaces and interfaces of microchips used in computers, and the heterogeneous catalysts found in the catalytic converter used for cleaning emissions in automobile exhausts. Surface chemistry also overlaps with electrochemistry. Surface science is of particular importance to the fields of heterogeneous catalysis, electrochemistry, and geochemistry.

Adsorption

Adsorption is the phenomenon that involves the attracting and retaining the molecules of a substance on the surface of a liquid or a solid which results in a higher concentration of molecules on the surface. This phenomenon is known as positive adsorption. The surface on which the adsorption occurs is called the adsorbent and the substance which is being adsorbed is called the adsorbate.

Types of Adsorption

Adsorption is of two types; Physical Adsorption (Physisorption) and Chemical Adsorption (Chemisorption). Let's understand them and their distinctions in detail.

Physical Adsorption (Physisorption)

Physisorption is a process in which the electronic structure of the atom or the molecule is barely perturbed upon adsorption. Physical adsorption occurs when a gas is adsorbed on the surface of a solid by Vander Waals Forces without forming any chemical bond between the adsorbate and the adsorbent.

Physical adsorption is used to characterize the surface and pore features of solids, that is, the material's texture. The adsorption isotherm generated by the Physical Adsorption technique enables one to characterize a solid's texture by determining its surface area, porosity, and total pore volume.

Chemical Adsorption (Chemisorption)

Chemisorption is a process which involves a chemical reaction between the surface and the adsorbate leading to the generation of chemical bonds at the adsorbent surface. Chemical adsorption occurs when a gas is adsorbed on the surface of a solid by forces identical to those of a chemical bond type of adsorption. Chemical adsorption leads to the formation of a surface compound. Chemisorption is irreversible and involves more heat than physical adsorption.

Chemisorption is used in the development, monitoring, and measuring of corrosion inhibitors and also for measuring surface active sites on a substrate.

Differences between Physisorption and Chemisorption

Physisorption	Chemisorption
The individuality of the adsorbate and the adsorbent are preserved.	There is a transfer/sharing of electrons or breakage of the adsorbate into atoms or radicals which are bound separately.
Only the Vander Waals Forces are present between the adsorbate and the surface of the adsorbent.	Chemical bonds are formed between the adsorbate and the surface of the adsorbent.
Usually, occurs at a low temperature and decreases with increasing temperature.	Occurs at a high temperature.
It is reversible.	It's irreversible.
Is related to the case of liquefaction of gases.	The extent of adsorption is not related to the liquefaction of the gas.
Forms multi-molecular layers.	Forms mono-molecular layers.
Does not require any activation energy.	Needs activation energy.

Although the theoretical difference physical and chemical adsorption is clear, the distinction is not as simple as it appears. You can use the following parameters to evaluate an adosrbate-adsorbent system to establish the type of adsorption:

- The heat of physical adsorption is in the same order of magnitude as the heat of condensation whereas the heat of chemical adsorption is of the same order as of the corresponding.
- Physical adsorption occurs under suitable temperature-pressure conditions in any gas- solid system while chemisorption takes place only if the gas is capable of forming a chemical bond with the surface.

- A physically adsorbed molecule can be removed unchanged at a reduced pressure at the same temperature at which the adsorption took place whereas the removal of the chemisorbed layer is much more difficult.
- Physical adsorption can involve the formation of multi-molecular layers and in some cases may take place on top of a chemisorbed monolayer. Chemisorption is always completed by the formation of a monolayer.
- Physical adsorption is instantaneous (although the diffusion into porous adsorbent may take some time) whereas even though chemisorption may be instantaneous, it generally needs activation energy.

Desorption

Desorption is the reverse of adsorption and involves the removal of the adsorbed substance from the surface. Desorption is the release of one substance from another, either from the surface or through the surface. Desorption can occur when an equilibrium situation is altered.

Desorption can be conducted by reducing the pressure and by heating.

Sorption

In absorption, the concentration of the molecules of a substance is largely in bulk than at the surface and the molecules in bulk are uniformly distributed. However, negative adsorption takes place if the concentration of the molecules of a substance is largely in bulk and the molecules are not uniformly distributed. The term sorption is used when both adsorption and absorption take place together and are not distinguishable.

Catalysis

The adhesion of gas or liquid molecules to the surface is known as adsorption. This can be due to either chemisorption or physisorption, and the strength of molecular adsorption to a catalyst surface is critically important to the catalyst's performance. However, it is difficult to study these phenomena in real catalyst particles, which have complex structures. Instead, well-defined single crystal surfaces of catalytically active materials such as platinum are often used as model catalysts. Multi-component material systems are used to study interactions between catalytically active metal particles and supporting oxides; these are produced by growing ultra-thin films or particles on a single crystal surface.

Relationships between the composition, structure, and chemical behavior of these surfaces are studied using ultra-high vacuum techniques, including adsorption and temperature-programmed desorption of molecules, scanning tunneling microscopy, low energy electron diffraction, and Auger electron spectroscopy. Results can be fed into chemical models or used toward the rational design of new catalysts. Reaction mechanisms can also be clarified due to the atomic-scale precision of surface science measurements.

The phenomenon of increase in the rate of reaction with the aid of a catalyst is called Catalysis. There are two types of Catalysis; Homogenous Catalysis and Heterogeneous Catalysis.

- Homogeneous Catalysis: This is when the catalyst and the reactants are in the same phase.
- Heterogeneous Catalysis: This is when the catalyst and the reactants are in different physical states.

Difference between Absorption and Adsorption

In the process of absorption, molecules of one phase interpenetrate uniformly among those of another phase to form a solution with the second phase. In adsorption, molecules of one phase are present in a higher concentration at the surface of the second phase. Water vapor is absorbed by calcium chloride while it is adsorbed by silica gel. In the same manner, ammonia is absorbed by water and is adsorbed by charcoal.

Simply put, adsorption is the adhesion of atoms, ions, or molecules from a gas, liquid, or dissolved solid to a surface. Absorption is the assimilation of molecules in the bulk phase (working principal of blotting paper). There are two types of adsorption:

- Physisorption: Adsorbate and adsorbent are held together by only van der Waal forces.
- Chemisorption: Adsorbate and adsorbent are held together by a chemical bond.

In Air-Conditioners, the outside air passes over activated charcoal having a larger surface area. The odorous chemicals get attached to the surface (physisorption) and thus the AC users get clean smelling air. During hydrogenation reaction for making Vanaspati ghee, the hydrogen gets adsorbed onto the nickel catalyst bed (chemisorption) and the H-H bond gets broken, this enables unsaturated oil molecules to get hydrogenated by the nascent H atom.

Electrochemistry

The behavior of an electrode-electrolyte interface is affected by the distribution of ions within the electrical double layer. Adsorption and desorption events can be studied at atomically flat single crystal surfaces as a function of applied bias, time, and solution conditions using scanning probe microscopy and surface X-ray scattering. These studies link traditional electrochemical techniques such as cyclic voltammetry to direct observations of interfacial processes.

Geochemistry

Geologic phenomena such as iron cycling and soil contamination are controlled by the interfaces between minerals and their environment. The atomic-scale structure and chemical properties of mineral-solution interfaces are studied using in situ synchrotron X-ray techniques such as X-ray reflectivity, X-ray standing waves, and X-ray absorption spectroscopy as well as scanning probe microscopy. For example, studies of heavy metal or actinide adsorption onto mineral surfaces reveal molecular-scale details of adsorption, enabling more accurate predictions of how these contaminants travel through soils or disrupt natural dissolution-precipitation cycles.

Colloids

In colloids, the dispersed substance aggregates into small particles. Then the surfaces of these particles interact with the dispersion media to form a stable dispersion. Soap molecules are a long non-polar organic chain terminating on one side with an ionic ion like a carboxylic acid group or triethylamine. While the ionic part is attracted to water, the non-polar organic part is repelled. To balance the repulsion and attraction, soap molecules come together and aggregate in the form of a ball where the outside surface of the ball is the ionic part and the inner portion is the organic chains. Any oil that comes into contact is picked up by the organic part and any metal salt is picked up by the ionic part. Thus, soap is used for cleaning. Colloidal solutions are different than true solutions and suspensions. Suspensions can be separated out through decantation.

References

- Wayne, C. E.; Wayne, R. P. Photochemistry, 1st ed.; Oxford University Press: Oxford, United Kingdom, reprinted 2005. ISBN 0-19-855886-4
- What-is-Physical-Chemistry, Chemistry: ivyroses.com, Retrieved 30 April 2018
- Headrick, J. M.; Diken, E. G.; Walters, R. S.; Hammer, N. I.; Christie, A.; Cui, J.; Myshakin, E. M.; Duncan, M. A.; Johnson, M. A.; Jordan, K. D. (2005). "Spectral Signatures of Hydrated Proton Vibrations in Water Clusters". Science. 308 (5729): 1765–69. Bibcode:2005Sci...308.1765H. doi:10.1126/science.1113094. PMID 15961665
- Physical-Chemistry: chemistryexplained.com, Retrieved 5 July 2018
- P. Klán, J. Wirz Photochemistry of Organic Compounds: From Concepts to Practice. Wiley, Chichester, 2009, ISBN 978-1405190886
- What-do-you-mean-by-femtochemistry: justscience.in, Retrieved 25 May 2018
- Schultz, M. J. (1999). "Why Equilibrium? Understanding the Role of Entropy of Mixing". J. Chem. Educ. 76 (10): 1391. Bibcode:1999JChEd..76.1391S. doi:10.1021/ed076p1391
- Thermodynamic-equilibrium, science: britannica.com, Retrieved 25 June 2018
- Blandamer, M. J. (1992). Chemical equilibria in solution: dependence of rate and equilibrium constants on temperature and pressure. New York: Ellis Horwood/PTR Prentice Hall. ISBN 0-13-131731-8

- Definition-of-equilibrium-constant-605099: thoughtco.com, Retrieved 11 July 2018
- Smiechowski, M.; Stangret, J. (2006). "Proton hydration in aqueous solution: Fourier transform infrared studies of HDO spectra". J. Chem. Phys. 125 (20): 204508–204522. Bibcode:-2006JChPh.125t4508S. doi:10.1063/1.2374891. PMID 17144716

2

Thermochemistry and Chemical Thermodynamics

Thermochemistry is a field of physical chemistry concerned with the study of heat energy in relation to chemical reactions and physical transformations. This chapter has been carefully written to provide an easy understanding of the varied aspects of chemical thermodynamics and thermochemistry, such as the fundamental principles of thermochemical equations, heat of combustion, exergonic and endergonic reactions, etc.

Thermochemistry

Thermochemistry is the study of the heat released or absorbed as a result of chemical reactions. It is a branch of thermodynamics and is utilized by a wide range of scientists and engineers. For example, biochemists use thermochemistry to understand bioenergetics, whereas chemical engineers apply thermochemistry to design manufacturing plants. Chemical reactions involve the conversion of a set of substances collectively referred to as "reactants" to a set of substances collectively referred to as "products." In the following balanced chemical reaction the reactants are gaseous methane, CH_4 (g), and gaseous molecular oxygen, O_2 (g), and the products are gaseous carbon dioxide, CO_2 (g), and liquid water H_2O(l):

$$CH_4(g) + 2O_2(g) \rightarrow CO_2(g) + 2H_2O(l)$$

Reactions in which a fuel combines with oxygen to produce water and carbon dioxide are called combustion reactions. Because natural gas consists primarily of methane, it is expected that above reaction will liberate heat. Reactions that liberate heat are termed exothermic reactions, and reactions that absorb heat are termed endothermic reactions.

The heat associated with a chemical reaction depends on the pressure and temperature at which the reaction is carried out. All thermochemical data presented here are for reactions carried out under standard conditions, which are a temperature of 298 K (24.85°C) and an applied pressure of one bar. The quantity of heat released in a reaction depends on the amount of material undergoing reaction. The chemical formulas that appear in a reaction each represent 1 mole of material; for example, the symbol CH_4 stands for 1 mole of methane having a mass of 16 grams (0.56 ounces), and the $2O_2$ (g) tells us that 2 moles of oxygen are required. Thermochemistry also depends on the physical state of the reactants and products. For example, the heat liberated in first equation is 890 kilojoules (kJ); if, however, water in the gas phase is formed, $H_2O(g)$, the heat released is only 802 kJ. Reversing a reaction like (l), which liberates heat, yields a reaction wherein heat must be supplied for the reaction to occur. The following reaction absorbs 890 kJ.

$$CO_2(g) + 2H_2O(l) \rightarrow CH_4(g) + 2O_2(g)$$

Energy and Enthalpy

Thermochemical changes are often discussed in terms of the "system" and the "surroundings." The system is regarded as the reaction products and reactants, whereas the surroundings consist of everything else in the universe. A boundary separates the system from the surroundings. The first law of thermodynamics relates the energy change belonging to a system to the amount of work and heat crossing the boundary. A statement of the first law applied to chemical reactions in which only heat and work cross the boundary is given by the expression:

$$U_{products} - U_{reactants} = \Delta U = q + w$$

Here U_{products} represents the energy of the products and $U_{\text{reactants}}$ represents the energy of the reactants. The heat associated with the reaction is given as q , and w represents work done during the transformation of reactants to products. If the volume of the system changes during the reaction and the applied pressure remains constant, the work carried out is termed pressure-volume work. For example, second reaction converts one mole of gas and two moles of liquid to a total of three moles of gas. The volume of the system increases during the reaction because, under standard conditions, a mole of gas occupies more volume than a mole of liquid. The work of expanding a system against atmospheric pressure is experienced when one inflates a balloon, and this work can be shown to be equal to $-P\Delta V$. Here P represents the atmospheric pressure and ΔV represents the change in volume of the system.

The first law of thermodynamics also states that U is a state function. State functions are very important in thermodynamics; they depend only on the present state of a system and not on its past history. Neither q nor w are state functions. An understanding of the concept of state function is furthered by considering the example of one's taking a trip from San Diego, California, to Denver, Colorado. The change in altitude that one experiences during this trip does not depend on the route taken and, thus, is similar to a state function. In comparison, the distance traveled between the two cities does depend on the route one follows; similarly, q and w are path-dependent quantities.

If a process such as a chemical reaction is carried out at a constant pressure in a way that involves only pressure-volume work, then $-P\Delta V$ can be substituted for the work term in third equation. Thus, we have:

$$\Delta U + P \Delta V = q_p$$

The symbol q_p represents the heat accompanying a chemical change carried out at constant pressure; in our previous example this would be equivalent to our specifying the exact route of travel between the two cities. The enthalpy of a system H is related to the energy of a system by the expression:

$$H = U + PV$$

For a process or reaction carried out at constant pressure:

$$\Delta H = \Delta U + P \Delta V = q_p$$

Enthalpy, like energy, is a state function. Thus, sixth equation shows that, for a reaction carried out at constant pressure, q_p depends only on the reactants consumed and the products formed. The enthalpy change associated with a reaction carried out under standard conditions is termed the heat of reaction and is given the symbol ΔH^o, with the superscript denoting standard conditions. Endothermic reactions have a positive ΔH^o whereas exothermic reactions have a negative ΔH^o. The change in enthalpy accompanying the conversion of reactants to products in a chemical reaction determines the amount of heat liberated or absorbed by the reaction. For a reaction carried out at constant pressure the enthalpy change depends only on the reactants and products.

Thermochemical Equations

An equation which indicates the evolution or absorption of heat in the reaction or process is called a thermochemical equation. For example, the following equation,

$$C + O_2 \rightarrow CO_2 + 393.5\text{kJ}$$

reveals that when carbon burns in oxygen to form carbon dioxide, 393.5kJ of heat are set free per one mole of carbon dioxide produced. Similarly the equation,

$$C + 2S \rightarrow CS_2 - 92.0\text{kJ}$$

reveals that one mole of carbon combines with 2 moles of sulfur to form one mole of carbon disulphide with the absorption of 92.0kJ of heat. Thus the equations written above are the thermochemical equations.

Laplace Law

A.L.Lavoisier and P.S.Laplace gave this law in 1780 which states that "the enthalpy of a reaction is exactly equal but opposite in sign for the reverse reaction."

For example, if DH is the enthalpy change in going from A to B then the enthalpy change for the process B to A would be -DH. Thus, the enthalpy of formation of a compound is numerically equal but opposite in sign to the enthalpy of decomposition of the compound.

$$S(s) + O_2(g) \rightarrow SO_2(g)\ \Delta H = -296.9\ kJ$$

$$SO_2(g) \rightarrow S(s) + O_2(g)\ \Delta H = +296.9\ kJ$$

Whenever a thermochemical equation is reversed the sign of DH also gets reversed.

Thermochemistry Problems

Below you could see problems.

Solved Examples

Question 1: Calculate the standard heat of formation of carbon disulphide (l). Given that the standard heats of combustion of carbon (s) sulfur (s) and carbon disulphide (l) are 393.3, -293.72 and -1108.76kJ mol^{-1}respectively.

Solution:

The given data can be written in thermochemical equation form as,

$$C_{(s)} + O_{2(g)} \rightarrow CO_{2(g)} \quad \Delta H = -393.5kJ$$

$$S_{(s)} + O_{2(g)} \rightarrow SO_{2(g)} \quad \Delta H = -293.72kJ$$

$$CS_2(1) + 3O_2 \rightarrow CO_{2(g)} + SO_{2(g)} \Delta H = -1108.76kJ$$

The required equation is,

$$C_{(s)} + 2S_{(g)} \rightarrow CS_{2(1)} \qquad \Delta H = ?$$

Multiplying second equation by 2 and adding to equation first we get,

$$2S_{(s)} + 2O_{2(g)} \rightarrow 2SO_{2(g)} \quad \Delta H = -587.44kJ$$

$$C_{(s)} + O_{2(g)} \rightarrow CO_{2(g)} \quad \Delta H = -393.5kJ$$

$$2S_{(s)} + C_{(s)} + O_{2(g)} \rightarrow 2SO_{2(g)} + CO_{2(g)}$$

Subtracting equation third from the above equation we have,

$$2S_{(s)} + C_{(s)} + O_{2(g)} \rightarrow 2SO_{2(g)} + CO_{2(g)} \quad \Delta H = -980.74kJ$$

$$CS_{2(l)} + 3O_{2(g)} \rightarrow CO_{2(g)} + SO_{2(g)} \quad \Delta H = -1108.76kJ$$

$$C_{(s)} + 2S_{(g)} \rightarrow CS_{2(1)} \qquad \Delta H = -128.02kJ$$

Question 2: Calculate lattice energy for the change,

$$Li^+(g) + CI^-(g) \rightarrow LiCI_{(s)}$$

Given that DH$_{subl}$. of Li = 160.67 kJ mol^{-1}, DH$_{Dissociation}$ of

Cl2 = 244.34 kJ mol^{-1}, DH$_{ionisation}$ of Li(g) = 520.07 kJ mol^{-1},

DH$_{E.A}$ of Cl(g) = - 365.26 kJ mol^{-1}, DH$^{o}_{f}$ of LiCl(s) = - 401.66 kJ mol^{-1}.

Solution:

Considering the different changes that occur in the formation of solid lithium chloride based on the data given the lattice energy of the above can be constituted as

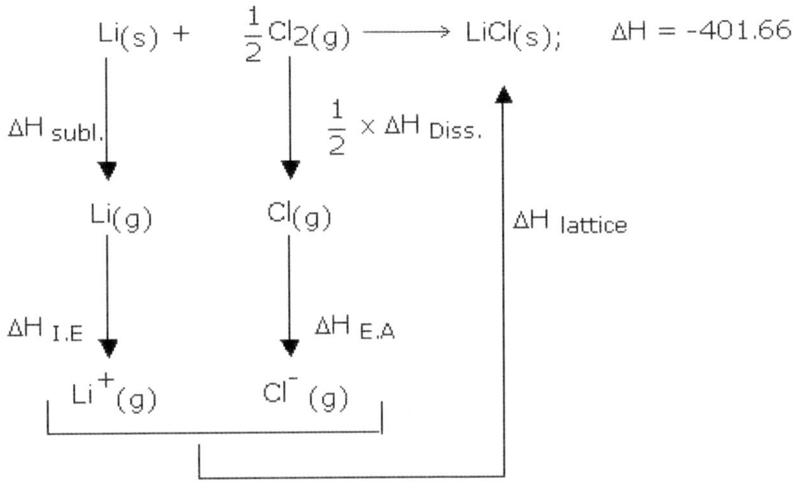

$$\Delta H^{o}_{f} = \Delta H_{subl.} + \Delta H_{I.E} + \frac{1}{2}\Delta H_{E.A} + \Delta H_{lattice}$$

Or,

$$\Delta H_{lattice} = \Delta H^{o}_{f} - \Delta H_{subl.} - \Delta H_{I.E} - \frac{1}{2}\Delta H_{Diss.}, - \Delta H_{E.A}$$

= - 839.31 kJ mol^{-1}

Hess's law

Every substance in this Universe, be it elements or compounds, have certain amount of energy called intrinsic energy. In a chemical reaction this energy gets re-distributed and is called the heat of reaction.

In a chemical reaction when the products formed have been found to have higher intrinsic energy, the reaction would require external supply of energy and so the reaction is termed as endothermic reaction. The products so formed have low intrinsic energy and the reaction itself is termed as exothermic reaction. The exothermic reactions release energy. Every chemical reaction involves a energy change in the form of heat. A reaction might proceed when this energy is supplied or the reactions are just spontaneous and release energy.

Usually the chemical reactions are run on constant pressure of 1 atmosphere and not on constant volume and hence the heat absorbed or evolved is found to be not equal to the intrinsic energy change (ΔE). The difference of this energy from the start of the reaction to the completion of the same is called its enthalpy change (ΔH). H can be thus related to E + PV. The difference in enthalpy is attributed to the change in volume as the pressure for the system remains constant.

Thus,

$$\Delta H = \Delta E + P\,\Delta V.$$

Hess's law is also known as Hess's law of constant heat summation. According to Hess's law, irrespective of the path taken by the reaction, the heat evolved or absorbed by the reaction at a constant pressure for any chemical change is same.

Consider a reaction $A \rightarrow C$ with enthalpy change as ΔH_1. Suppose the reaction proceeds in two steps $A \rightarrow B$ with enthalpy change as ΔH_2 and $B \rightarrow C$ with enthalpy change as ΔH_3. According to Hess's law ΔH_1 is the sum of ΔH_2 and ΔH_3.

Equation

$$\Delta H_1 = \Delta H_2 + \Delta H_3$$

This is Hess's Law equation.

The event of the reaction can be explained in more than two steps, the sum of the enthalpy changes steps involved in the complete reaction of reactants to form products give the enthalpy of reaction of the process.

During the course of a reaction some of the steps are found to be endothermic and the rest are exothermic in nature at constant pressure. The total enthalpy change for the reaction is nothing but the sum of all the steps for exothermic enthalpy change minus the sum of all steps of endothermic enthalpy change.

Hess's Law Examples

Let us now see an example with the values of ΔH° for each step. Generally, the values of ΔH° are given at room temperature or 298K. Carbon in the form of graphite burns completely in excess of oxygen to form carbon-di-oxide releasing heat.

$$C_{(graphite)} + O_{2(g)} \leftrightarrow CO_{2(g)}\ \Delta H_1^{o}{}_{(298K)} = -393.51\ kJ/mole.$$

Carbon monoxide also burns in oxygen to give Carbon dioxide releasing heat.

$$CO_{(g)} + \tfrac{1}{2}\,O_{2(g)} \leftrightarrow CO_{2(g)}\ \Delta H_2^{o}{}_{(298K)} = -282.98\ kJ/mole.$$

It is difficult to measure exactly the enthalpy change for the reaction,

$C_{(graphite)} + ½ O2_{(g)} \leftrightarrow CO_{(g)} .\Delta H3^o = ?$

However using Hess's equation one can find out the value.

$\Delta H_1^o - \Delta H_2^o = \Delta H_3^o$

Thus (-393.51 kJ/mole.) – (-282.98 kJ/mole) = – 110.53 kJ/mole.

Problems

Problem 1:

The heat of formations of methane, water(l) and carbon dioxide are -74.89 kJ mol^{-1}, 286.18 kJ mol-1and 393.51 kJ mol^{-1} respectively. Calculate the heat of combustion of methane at room temperature.

Solution: The given values are;

$C(s) + 2H2_{(g)} \rightarrow CH_{4(g)}$; ΔH = -74.89 kJ mol^{-1}

$H_{2(g)} + ½ O_{2(g)} \rightarrow H_2O_{(l)}$; ΔH = -286.18 kJ mol^{-1}

$C_{(graphite)} + O_{2(g)} \rightarrow CO_{2(g)}$ $\Delta H1^o_{(298K)}$ = -393.51 kJ mol_{-1}.

In order to get the value for $CH_4(g + 2 O_{2(g)} \rightarrow CO_{2(g)} + 2 H_2O_{(l)}$ we have to reverse the above first equation, take the third equation as it is and multiply second equation by 2. Thus,

$CH_{4(graphite)} \rightarrow C_{(s)} + 2H_{2(g)}$ = 74.89 kJ mol^{-1}

$C_{(graphite)} + O_{2(g)} \rightarrow CO_{2(g)}$ $\Delta H1^o_{(298K)}$ = -393.51 kJ mol^{-1}

2 X ($H_{2(g)} + ½ O_{2(g)} \rightarrow H_2O_{(l)}$) ; ΔH = 2X -286.18 kJ mol^{-1}

Therefore,

$CH_{4(g)} + 2 O_{2(g)} \rightarrow CO_{2(g)} + 2 H_2O_{(l)}$

Heat of combustion = 890 98 kJ mol^{-1}

Problem 2:

The heats of combustion of ethylene, hydrogen and ethane are -336.97 k cal, - 68.41 k cal and -337 k cal respectively at 298K. calculate the heat of reaction in the saturation of ethylene to ethane.

Solution: The given data is

$C_2H_{4(g)} + 3O_{2(g)} \rightarrow 2CO_{2\,(g)} + 2\,H_2O$ ΔH - 336.97 k cal.

$H_{2\,(g)} + ½ O_{2\,(g)} \rightarrow H_2O_{\,(l)}$ ΔH - 68.41 k cal

$$C_2H_{6\,(g)} + 3\tfrac{1}{2}\,O_{2\,(g)} \rightarrow 2CO_{2(g)} + 3\,H_2O_{(l)}\ \Delta H\ -373\ \text{k cal}$$

In order to get the heat of reaction above first and second equations are to be added. The value of ΔH in third equation in reverse becomes 373 k cal. which is to be subtracted.

Thus,

$$(-\,336.97) + (-\,68.41) = -\,405.38 \text{ and } (-\,405.38\,) - 373 = 32.38 \text{ k cal.}$$

Extension to Free Energy and Entropy

The concepts of Hess's law can be expanded to include changes in entropy and in Gibbs free energy, which are also state functions. The Bordwell thermodynamic cycle is an example of such an extension which takes advantage of easily measured equilibria and redox potentials to determine experimentally inaccessible Gibbs free energy values. Combining ΔG° values from Bordwell thermodynamic cycles and ΔH° values found with Hess's law can be helpful in determining entropy values which are not measured directly, and therefore must be calculated through alternative paths.

For the free energy:

$$\Delta G^{\ominus}_{reaction} = \sum \Delta G^{\ominus}_{\mathrm{f}(products)} - \sum \Delta G^{\ominus}_{\mathrm{f}(reactants)}.$$

For entropy, the situation is a little different. Because entropy can be measured as an absolute value, not relative to those of the elements in their reference states (as with ΔH° and ΔG°), there is no need to use the entropy of formation; one simply uses the absolute entropies for products and reactants:

$$\Delta S^{\ominus}_{reaction} = \sum S^{\ominus}_{(products)} - \sum S^{\ominus}_{(reactants)}.$$

Application

Hess's law has been useful in determining the heat changes of reactions, which cannot be measured directly with a calorimeter. Some of its applications are:

1. Determination of Heat of Formation

Compounds whose heats of formation cannot be measured directly using colorimetric methods because they cannot be synthesized from their elements easily e.g. methane, carbon monoxide, benzene etc are determined using Hess's Law.

For example, the heat of formation of carbon monoxide can be calculated from the heat of combustion data for carbon and carbon monoxide.

2. Determination of Heat of Transition

The heats of transition of allotropic modification of compounds such as diamond to graphite, rhombic sulphur to monoclinic sulfur, yellow phosphorous to red phosphorous etc. can be determined using Hess's Law.

For example, the heat of transition of diamond to graphite can be calculated from the heat of combustion data for diamond and graphite, which is -395.4 kJ and -393.5 kJ respectively.

The thermochemical equations showing the combustion reaction of diamond and graphite are

$$C_{(diamond)} + O_{2(g)} \rightarrow CO_{2\,(g)} \qquad \Delta H = -395.4 kJ$$
$$C_{(graphite)} + O_{2(g)} \rightarrow CO_{2\,(g)} \qquad \Delta H = -393.5 kJ$$

The conversion that is required is

$$C_{(diamond)} \rightarrow C_{(graphite)} \qquad \Delta H = ?$$

This can be obtained by subtracting the second equation from the first one.

$$C_{(diamond)} + O_{2(g)} \rightarrow CO_{2\,(g)} \qquad \Delta H = -395.4 kJ$$
$$C_{(graphite)} + O_{2(g)} \rightarrow CO_{2\,(g)} \qquad \Delta H = -393.5 kJ$$
$$C_{(diamond)} \rightarrow C_{(graphite)} \qquad \Delta H = -1.9 kJ$$

3. Determination of Heat of Formation

The heat of hydration of substances is calculated using Hess's law.

For example the heat of hydration of copper sulphate can be calculated from the heat of solution of anhydrous and hydrated salts of copper. The heat of solution of $CuSO_4$ and $CuSO_4.5H_2O$ are -66.5 and -11.7 kJ mol^{-1}. The corresponding thermochemical equations are:

$$CuSO_{4(S)} + aq \rightarrow CuSO_{4(aq)} \quad \Delta H_1 = -66.5 kJ$$
$$CuSO_4.5H_2O_{(S)} + aq \rightarrow CuSO_{4(aq)} \quad \Delta H_2 = -11.7 kJ$$

The process of hydration can be expressed as,

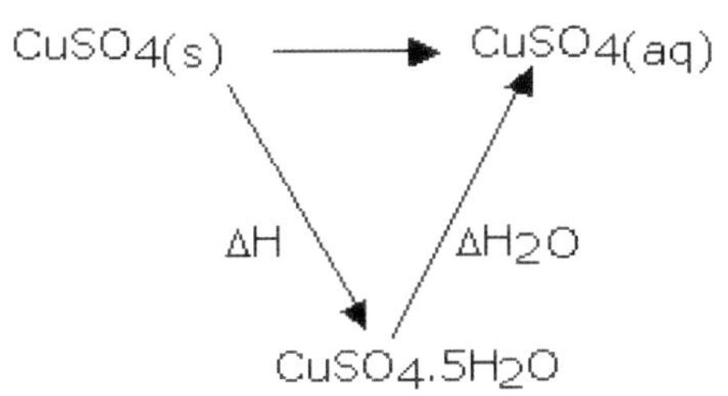

According to Hess law, $DH_1 = DH + DH_2$,

$$DH = DH_1 - DH_2$$

$$= - 66.5\ 11.7 = - 78.2\ kJ/mol$$

4. Determination of Heat of Various Reactions

Hess's law is useful in calculating the enthalpies of many reactions where direct measurement is difficult or impossible.

Chemical Thermodynamics

Chemical thermodynamics is the study of how heat and work relate to each other both in changes of state and in chemical reactions. It involves a series of rules and laws that explain how heat and work well, work, and explains which processes can happen spontaneously and which need some help.

There are several basic principles of chemical thermodynamics to consider: systems, the laws of thermodynamics, and enthalpy. Chemical thermodynamics is also concerned with four particular quantities: internal energy, enthalpy, entropy and the Gibbs free energy.

Systems in Chemical Thermodynamics

To understand thermodynamics, it's helpful to first define something called a system. A system is a series of components that are connected together. In a nutshell, it's the part of the world we're focusing on. We can look at what moves in and out of a particular system. For example, if we have coffee in a thermos flask, we can either call the coffee itself the system, or the whole thermos flask (including the walls of the container).

There are several types of systems in chemical thermodynamics: isolated systems, closed systems, and open systems.

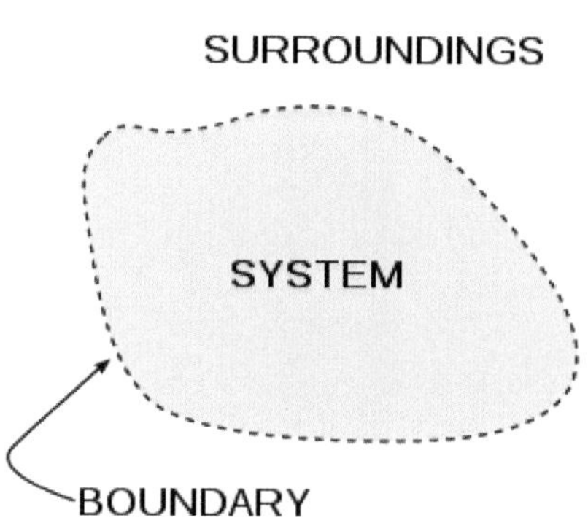

The boundary of a system defines it

An isolated system is one that has rigid walls and doesn't allow the transfer of energy or mass. The walls are perfectly insulating. A closed system has walls that let energy pass in and out of the system, but that don't allow mass to enter or escape. And an open system allows *both* energy and matter to enter and leave.

Laws of Thermodynamics

There are several important laws of thermodynamics, which form the cornerstone of the field. The first law of thermodynamics says that the change in the internal energy of a closed system equals the heat added to the system, minus the work done by the system on the surroundings. This also means that in an isolated system, the energy changes inside the system must be zero.

The second law of thermodynamics says that the entropy in the universe must always increase. Entropy is the amount of disorder in the universe, measured in joules per kelvin. So this means the universe always becomes more disorderly. Let's say that you decide to tidy your home: you put away the books in alphabetical order, wash the dishes and store them in neat piles. You might think you've just made the universe more orderly, and in some ways you have, but in the universe as a whole this is impossible. By using your muscles to tidy things up, you've produced heat energy in your arms, and that heat energy has overall made the universe *less* orderly.

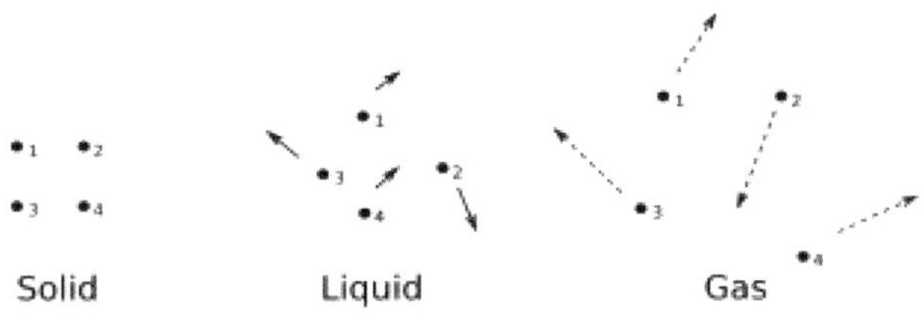

Gases are more disorderly than liquids and solids, and so have a greater entropy

Another consequence of the second law of thermodynamics is that heat can only travel spontaneously from hot places to cold places. This means that a refrigerator can't work on its own. The only way to make a refrigerator work is by doing work - by using energy from the electricity supply in the wall. Turn off the electricity, and heat will move from hot places (the room) to cold places (the refrigerator) just like normal.

Chemical Reactions

In most cases of interest in chemical thermodynamics there are internal degrees of freedom and processes, such as chemical reactions and phase transitions, which always create entropy unless they are at equilibrium, or are maintained at a "running equilibrium" through "quasi-static" changes by being coupled to constraining devices, such as pistons or electrodes, to deliver and receive external work. Even for homogeneous "bulk" materials, the free energy functions depend on the composition, as do all the extensive thermodynamic potentials, including the internal energy. If the quantities

$\{N_i\}$, the number of chemical species, are omitted from the formulae, it is impossible to describe compositional changes.

Gibbs Function or Gibbs Energy

For a "bulk" (unstructured) system they are the last remaining extensive variables. For an unstructured, homogeneous "bulk" system, there are still various *extensive* compositional variables $\{N_i\}$ that G depends on, which specify the composition, the amounts of each chemical substance, expressed as the numbers of molecules present or (dividing by Avogadro's number = 6.023 × 10^{23}), the numbers of moles,

$$G = G(T, P, \{N_i\}).$$

For the case where only PV work is possible,

$$\mathrm{d}G = -S\mathrm{d}T + V\mathrm{d}P + \sum_i \mu_i \mathrm{d}N_i$$

in which μ_i is the chemical potential for the i-th component in the system,

$$\mu_i = \left(\frac{\partial G}{\partial N_i}\right)_{T,P,N_{j\neq i},etc.}.$$

The expression for dG is especially useful at constant T and P, conditions which are easy to achieve experimentally and which approximates the condition in living creatures,

$$(\mathrm{d}G)_{T,P} = \sum_i \mu_i \mathrm{d}N_i.$$

Chemical Affinity

While this formulation is mathematically defensible, it is not particularly transparent since one does not simply add or remove molecules from a system. There is always a *process* involved in changing the composition; e.g., a chemical reaction (or many), or movement of molecules from one phase (liquid) to another (gas or solid). We should find a notation which does not seem to imply that the amounts of the components (N_i) can be changed independently. All real processes obey conservation of mass, and in addition, conservation of the numbers of atoms of each kind. Whatever molecules are transferred to or from should be considered part of the "system".

Consequently, we introduce an explicit variable to represent the degree of advancement of a process, a progress variable ξ for the *extent of reaction* and to the use of the partial derivative $\partial G/\partial\xi$ (in place of the widely used "ΔG", since the quantity at issue is not a finite change). The result is an understandable expression for the dependence of dG on chemical reactions (or other processes). If there is just one reaction,

$$(dG)_{T,P} = \left(\frac{\partial G}{\partial \xi}\right)_{T,P} d\xi.$$

If we introduce the *stoichiometric coefficient* for the *i-th* component in the reaction,

$$\nu_i = \partial N_i / \partial \xi$$

which tells how many molecules of *i* are produced or consumed, we obtain an algebraic expression for the partial derivative,

$$\left(\frac{G}{\partial \xi}\right)_{T,P} = \sum_i \mu_i \nu_i = -\mathbb{A}$$

where, we introduce a concise and historical name for this quantity, the "affinity", symbolized by A, as introduced by Théophile de Donder in 1923. The minus sign comes from the fact the affinity was defined to represent the rule that spontaneous changes will ensue only when the change in the Gibbs free energy of the process is negative, meaning that the chemical species have a positive affinity for each other. The differential for *G* takes on a simple form which displays its dependence on compositional change,

$$(dG)_{T,P} = -\mathbb{A}\, d\xi.$$

If there are a number of chemical reactions going on simultaneously, as is usually the case,

$$(dG)_{T,P} = -\sum_k \mathbb{A}_k\, d\xi_k.$$

a set of reaction coordinates $\{\xi_j\}$, avoiding the notion that the amounts of the components (N_i) can be changed independently. The expressions above are equal to zero at thermodynamic equilibrium, while in the general case for real systems, they are negative because all chemical reactions proceeding at a finite rate produce entropy. This can be made even more explicit by introducing the reaction *rates* $d\xi_j/dt$. For each and every *physically independent process*,

$$\mathbb{A}\,\dot{\xi} \le 0.$$

This is a remarkable result since the chemical potentials are intensive system variables, depending only on the local molecular milieu. They cannot "know" whether the temperature and pressure (or any other system variables) are going to be held constant over time. It is a purely local criterion and must hold regardless of any such constraints. Of course, it could have been obtained by taking partial derivatives of any of the other fundamental state functions, but nonetheless is a general criterion for ($-T$ times) the entropy production from that spontaneous process; or at least any part of it that is not captured as external work.

We now relax the requirement of a homogeneous "bulk" system by letting the chemical potentials and the affinity apply to any locality in which a chemical reaction (or any other process) is occurring. By accounting for the entropy production due to irreversible processes, the inequality for dG is now replaced by an equality,

$$\mathrm{d}G = -S\mathrm{d}T + V\mathrm{d}P - \sum_k \mathbb{A}_k \mathrm{d}\xi_k + W'$$

or

$$\mathrm{d}G_{T,P} = -\sum_k \mathbb{A}_k \mathrm{d}\xi_k + W'.$$

Any decrease in the Gibbs function of a system is the upper limit for any isothermal, isobaric work that can be captured in the surroundings, or it may simply be dissipated, appearing as T times a corresponding increase in the entropy of the system and/or its surrounding. Or it may go partly toward doing external work and partly toward creating entropy. The important point is that the *extent of reaction* for a chemical reaction may be coupled to the displacement of some external mechanical or electrical quantity in such a way that one can advance only if the other one also does. The coupling may occasionally be *rigid*, but it is often flexible and variable.

Solutions

In solution chemistry and biochemistry, the Gibbs free energy decrease ($\partial G/\partial\xi$, in molar units, denoted cryptically by ΔG) is commonly used as a surrogate for ($-T$ times) the entropy produced by spontaneous chemical reactions in situations where there is no work being done; or at least no "useful" work; i.e., other than perhaps some $\pm P\,\mathrm{d}V$. The assertion that all *spontaneous reactions have a negative* ΔG *is merely a restatement of the fundamental thermodynamic relation, giving it the physical dimensions of energy and somewhat obscuring its significance in terms of entropy. When there is no useful work being done, it would be less misleading to use the Legendre transforms of the entropy appropriate for constant* T, *or for constant* T *and* P, *the Massieu functions* −F/T *and* −G/T *respectively.*

Non Equilibrium

Generally the systems treated with the conventional chemical thermodynamics are either at equilibrium or near equilibrium. Ilya Prigogine developed the thermodynamic treatment of open systems that are far from equilibrium. In doing so he has discovered phenomena and structures of completely new and completely unexpected types. His generalized, nonlinear and irreversible thermodynamics has found surprising applications in a wide variety of fields.

The non equilibrium thermodynamics has been applied for explaining how ordered structures e.g. the biological systems, can develop from disorder. Even if Onsager's relations are utilized, the classical principles of equilibrium in thermodynamics still show that linear systems close to equilibrium always develop into states of disorder which are stable to perturbations and cannot explain the occurrence of ordered structures.

Prigogine called these systems dissipative systems, because they are formed and maintained by the dissipative processes which take place because of the exchange of energy between the system and its environment and because they disappear if that exchange ceases. They may be said to live in symbiosis with their environment.

The method which Prigogine used to study the stability of the dissipative structures to perturbations is of very great general interest. It makes it possible to study the most varied problems, such as city traffic problems, the stability of insect communities, the development of ordered biological structures and the growth of cancer cells to mention but a few examples.

System Constraints

In this regard, it is crucial to understand the role of walls and other *constraints*, and the distinction between *independent* processes and *coupling*. Contrary to the clear implications of many reference sources, the previous analysis is not restricted to homogeneous, isotropic bulk systems which can deliver only PdV work to the outside world, but applies even to the most structured systems. There are complex systems with many chemical "reactions" going on at the same time, some of which are really only parts of the same, overall process. An *independent* process is one that *could* proceed even if all others were unaccountably stopped in their tracks. Understanding this is perhaps a "thought experiment" in chemical kinetics, but actual examples exist.

A gas reaction which results in an increase in the number of molecules will lead to an increase in volume at constant external pressure. If it occurs inside a cylinder closed with a piston, the equilibrated reaction can proceed only by doing work against an external force on the piston. The extent variable for the reaction can increase only if the piston moves, and conversely, if the piston is pushed inward, the reaction is driven backwards.

Similarly, a redox reaction might occur in an electrochemical cell with the passage of current in wires connecting the electrodes. The half-cell reactions at the electrodes are constrained if no current is allowed to flow. The current might be dissipated as joule heating, or it might in turn run an electrical device like a motor doing mechanical work. An automobile lead-acid battery can be recharged, driving the chemical reaction backwards. In this case as well, the reaction is not an independent process. Some, perhaps most, of the Gibbs free energy of reaction may be delivered as external work.

The hydrolysis of ATP to ADP and phosphate can drive the force times distance work delivered by living muscles, and synthesis of ATP is in turn driven by a redox chain in mitochondria and chloroplasts, which involves the transport of ions across the membranes of these cellular organelles. The coupling of processes here, and in the previous examples, is often not complete. Gas can leak slowly past a piston, just as it can slowly leak out of a rubber balloon. Some reaction may occur in a battery even if no external current is flowing. There is usually a coupling coefficient, which may depend on relative rates, which determines what percentage of the driving free energy is turned into external work, or captured as "chemical work"; a misnomer for the free energy of another chemical process.

Exothermic Process

Exothermic Reactions

An exothermic reaction (the thermite reaction) using Iron (III) Oxide. The sparks flying outwards are drops of molten iron trailing

An exothermic reaction occurs when the energy used to break the bonds in the reactants (the starting stuff) is less than the energy released when new bonds are made in the products (the stuff you end up with). This extra energy is given off as heat and there is a temperature rise around the surroundings of the reaction.

Combustion is an example of an exothermic reaction.

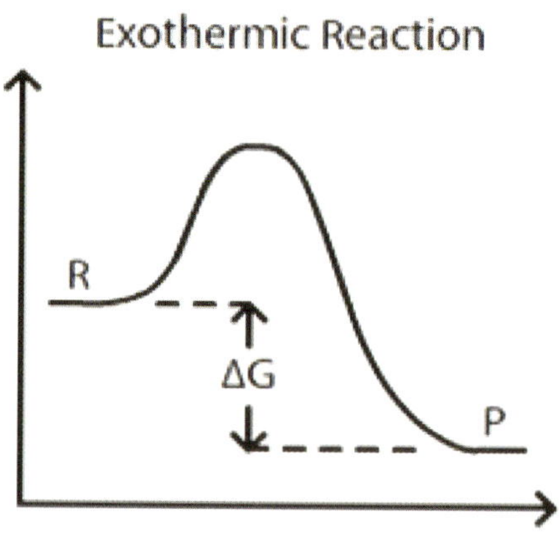

This graph shows that energy has been released and Delta H (energy change) is negative.

The reactants have more energy than the end products. It also indicates that the enthalpy change is negative as heat is lost to the surroundings resulting in a temperature increase.

Endothermic Process

Endothermic Reactions

An endothermic reaction occurs when the energy used to break the bonds in the reactants is greater than the energy given out when bonds are formed in the products. This means that overall the reaction takes in energy, therefore there is a temperature decrease in the surroundings.

Electrolysis is an example of an endothermic reaction but you can create one easily in the kitchen just by dissolving salt or sugar in water.

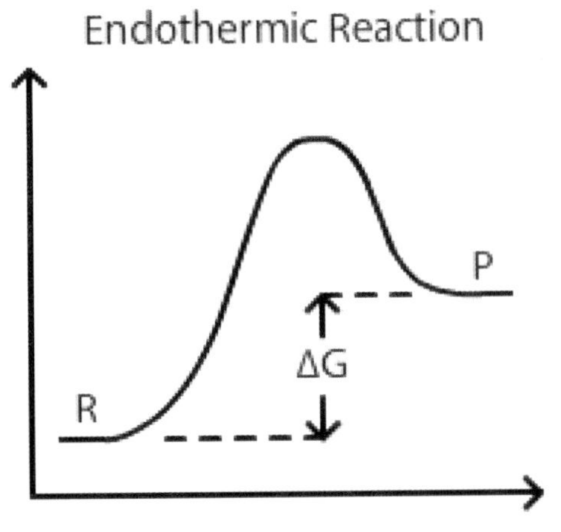

This graph shows that energy has been taken in and Delta H is positive.

Similarly, this endothermic chart shows that the reactants start off with less energy this time and the end products have absorbed energy. Clearly shown in the diagram, enthalpy change of the reaction is positive because energy is being taken in from the surroundings which results in a temperature decrease.

Heat of Combustion

The heat of combustion is the energy liberated when a substance undergoes complete combustion, at constant pressure usually in an environment with excess Oxygen. The heat of combustion is utilised to quantify the performance of a fuel in a combustion system such as furnaces, power generation turbines and motors. The heat of combustion and provides a list of heats of combustion for commons fuels and fuel components are described here.

Heating Value

The heat of combustion is typically presented in the form of a heating value. The heating value is the amount of energy released during combustion and can be referenced as a higher or lower heating value.

Higher Heating Value

The higher heating value (HHV) accounts for the heat of combustion and any energy released to bring the combustion products back their pre-combustion temperatures (typically 25°C). By bringing the combustion products back to pre-combustion temperatures, the water component of the combustion products condenses and therefore the latent heat of vaporisation of water is incorporated in the higher heating value. The higher heating value is most useful in circumstances where condensation of combustion products is practical.

Lower Heating Value

The lower heating value (LHV) assumes that the combustion products are not brought back to pre-combustion temperatures and is therefore essentially the higher heating value minus the latent heat of vaporisation of the water product. The lower heating value may be approximated from the higher heating value as follows:

$$LHV = HHV - h_{vap,water} \times \frac{nH2O}{n\,fuel}$$

The accuracy of the conversion between LHV and HHV may be improved by accounting for the change in sensible heat of the combustion products.

Gross Heating Value

Gross heating value accounts for water in the exhaust leaving as vapor, and includes liquid water in the fuel prior to combustion. This value is important for fuels like wood or coal, which will usually contain some amount of water prior to burning.

Measuring Heating Values

The higher heating value is experimentally determined in a bomb calorimeter. The combustion of a stoichiometric mixture of fuel and oxidizer (e.g. two moles of hydrogen and one mole of oxygen) in a steel container at 25 °C (77 °F) is initiated by an ignition device and the reactions allowed to complete. When hydrogen and oxygen react during combustion, water vapor is produced. The vessel and its contents are then cooled to the original 25°C and the higher heating value is determined as the heat released between identical initial and final temperatures.

When the lower heating value (LHV) is determined, cooling is stopped at 150°C and the reaction heat is only partially recovered. The limit of 150°C is based on acid gas dew-point.

Note: Higher heating value (HHV) is calculated with the product of water being in liquid form while lower heating value (LHV) is calculated with the product of water being in vapor form.

Relation between Heating Values

The difference between the two heating values depends on the chemical composition of the fuel. In the case of pure carbon or carbon monoxide, the two heating values are almost identical, the difference being the sensible heat content of carbon dioxide between 150°C and 25°C (sensible heat exchange causes a change of temperature. In contrast, latent heat is added or subtracted for phase transitions at constant temperature. Examples: heat of vaporization or heat of fusion). For hydrogen the difference is much more significant as it includes the sensible heat of water vapor between 150°C and 100°C, the latent heat of condensation at 100°C, and the sensible heat of the condensed water between 100°C and 25°C. All in all, the higher heating value of hydrogen is 18.2% above its lower heating value (142 MJ/kg vs. 120 MJ/kg). For hydrocarbons the difference depends on the hydrogen content of the fuel. For gasoline and diesel the higher heating value exceeds the lower heating value by about 10% and 7% respectively, and for natural gas about 11%.

A common method of relating HHV to LHV is:

$$\mathrm{HHV} = \mathrm{LHV} + H_{\mathrm{v}}\left(\frac{n_{\mathrm{H_2O,out}}}{n_{\mathrm{fuel,in}}}\right)$$

where H_v is the heat of vaporization of water, $n_{H2O,out}$ is the moles of water vaporized and $n_{fuel,in}$ is the number of moles of fuel combusted.

- Most applications that burn fuel produce water vapor, which is unused and thus wastes its heat content. In such applications, the lower heating value must be used to give a 'benchmark' for the process.

- However, for true energy calculations in some specific cases, the higher heating value is correct. This is particularly relevant for natural gas, whose high hydrogen content produces much water, when it is burned in condensing boilers and power plants with flue-gas condensation that condense the water vapor produced by combustion, recovering heat which would otherwise be wasted.

Usage of Terms

Many engine manufacturers rate their engine fuel consumption by the lower heating values. American consumers should be aware that the corresponding fuel-consumption figure based on the higher heating value will be somewhat higher.

The difference between HHV and LHV definitions causes endless confusion when quoters do not bother to state the convention being used. since there is typically a 10% difference between the two methods for a power plant burning natural gas. For simply benchmarking part of a reaction the LHV may be appropriate, but HHV should be used for overall energy efficiency calculations, if only to avoid confusion, and in any case the value or convention should be clearly stated.

Accounting for Moisture

Both HHV and LHV can be expressed in terms of AR (all moisture counted), MF and MAF (only water from combustion of hydrogen). AR, MF, and MAF are commonly used for indicating the heating values of coal:

- AR (as received) indicates that the fuel heating value has been measured with all moisture- and ash-forming minerals present.
- MF (moisture-free) or dry indicates that the fuel heating value has been measured after the fuel has been dried of all inherent moisture but still retaining its ash-forming minerals.
- MAF (moisture- and ash-free) or DAF (dry and ash-free) indicates that the fuel heating value has been measured in the absence of inherent moisture- and ash-forming minerals.

Experimental Calculation of Heat of Combustion

In the absence of published data the heat of combustion (LHV) can be experimentally determined using the following procedure:

1. Measure a known quantity of water into a flask and stand this on top of a tripod situated above a fuel source.
2. Measure the initial temperature of the water and keep a thermometer submersed in the water to measure any temperature changes.
3. Measure the initial mass of the fuel being used. For example methanol in a controlled burning manner e.g. through a wick.
4. When the temperature of the water has increase by at least 10 C then the experiment can be stopped by extinguishing the flame.
5. Record the final mass of the fuel.

6. Record the final temperature of the water and calculate the temperature change.
7. Calculate the heat of combustion using the methodology demonstrated below.

An example of the experimental calculation process is shown as follows. Please note that these numbers are not obtained from an actual experiment, they are purely for demonstrating the experimental and calculation method to apply if performing the heat of combustion determination experiment please refer to literature for actual heat of combustion values for different fuels and do not undertake any experiments without completely understanding and controlling the hazards associated with the experiment.

Initial temperature of water	20°C
Final temperature of water	30°C
Change in temperature of water	10°C
Mass of water	100g
Initial mass of methanol	20.00g
Final mass of methanol	19.79g
Fuel consumed	0.21g

1. Calculate mols of fuel consumed:

MW of methanol = 32 g/mol

mols (n) = 0.21 g / 32 g/mol = 6.55×10^{-3} mol of methanol

2. Calculate enthalpy increase of water using $Q=mc_p\Delta T$:

Mass of water = 100 g

Specific heat = 4.18 J/g.K

Temperature change = 10 °C (K)

Enthalpy = 100g x 4.18 J/g.K x 10 K = 4180J = 4.18 kJ

3. Calculate the heat of combustion assuming heat not lost to surroundings: 6.55×10^{-3} mol of methanol produces 4.18kJ of heat.

4.18 kJ / 6.55×10^{-3} mol = 638.6 kJ/mol = experimental ΔH_C kJ/mol methanol.

Thermochemical Equation

A thermochemical equation has two parts: a balanced chemical equation and the change in one or more thermodynamic quantities (*e.g.*, temperature, energy, or enthalpy) that occurs when that change occurs. The balanced equation can describe either a physical change or a chemical change. In the example above, the equation states that when one mole of solid water melts into liquid water at 0°C, the enthalpy of the system increases by 6.00 kJ (the reactants and products in a thermochemical equation will almost always be the system in a thermodynamics problem). This module will show you how thermochemical equations work and how a relatively small number of special thermochemical equations allow you to predict the outcomes of many other reactions.

Energy changes which accompany chemical reactions are almost always expressed by thermochemical equations, such as

$CH_4(g)+2O_2(g)\rightarrow CO_2(g)+2H_2O(l)$ (25°C, 1 atm pressure) $\Delta H_m=-890kJ$

which is displayed on the atomic level below. To get an idea of what this reaction looks like on the macroscopic level, check out the flames on the far right.

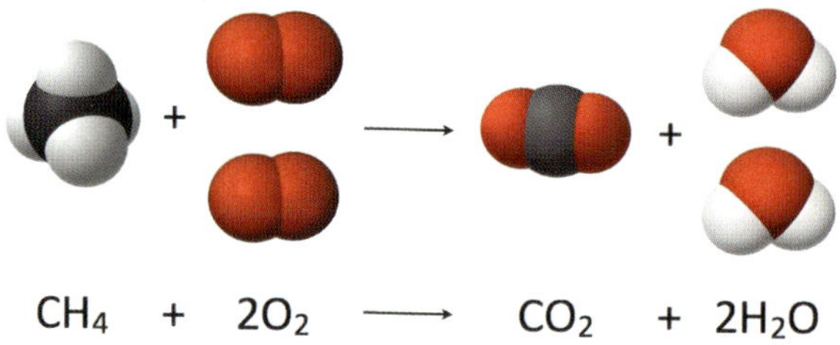

Here the ΔH_m (delta *H* subscript m) tells us whether heat energy is released or absorbed when the reaction occurs as written, and also enables us to find the actual quantity of energy involved. By convention, if ΔH_m is *positive*, heat is *absorbed* by the reaction; i.e., it is endothermic. More commonly, ΔH_m is *negative* as in above equation indicating that heat energy is *released* rather than absorbed by the reaction, and that the reaction is exothermic. This convention as to whether ΔH_m is positive or negative looks at the heat change in terms of the matter actually involved in the reaction rather than its surroundings. In the reaction in above equation the C, H, and O atoms have collectively lost energy and it is this loss which is indicated by a negative value of ΔH_m.

It is important to notice that ΔH_m is the energy for the reaction as written. In the case of above equation that represents the formation of 1 mol of carbon dioxide and 2 mol of water. The quantity of heat released or absorbed by a reaction is proportional to the amount of each substance consumed or produced by the reaction. Thus above equation tells us that 890.4 kJ of heat energy is given off *for every mole of* CH_4 which is consumed. Alternatively, it tells us that 890.4 kJ is released *for every 2 moles of* H_2O produced. Seen in this way, ΔH_m is a conversion factor enabling us to calculate the heat absorbed or released when a given amount of substance is consumed or produced. If q is the quantity of heat absorbed or released and n is the amount of substance involved, then

$$\Delta H_m = \frac{q}{n}$$

Example: Heat Energy

How much heat energy is obtained when 1 kg of ethane gas, C_2H_6, is burned in oxygen according to the equation:

$$2C_2H_6(g)+7O_2(g)\rightarrow 4CO_2(g)+6H_2O(l)\ \Delta Hm=-3120\ kJ$$

Solution: The mass of C_2H_6 is easily converted to the amount of C_2H_6 from which the heat energy q is easily calculated by means of second equation. The value of ΔH_m is −3120 kJ per per 2 mol C_2H_6. The road map is

$$mC_2H_6 \underrightarrow{M} nC_2H_6 \underrightarrow{\Delta H_m} q$$

so that

$$q = 1\times 10^3\ g\,C_2H_6 \times \frac{1\ mol\,C_2H_6}{30.07\ g\,C_2H_6} \times \frac{-3120\,kJ}{2\ mol\,C_2H_6}$$
$$= -51\ 879\ kJ = -51.88\ MJ$$

By convention, a negative value of q corresponds to a release of heat energy by the matter involved in the reaction.

The quantity ΔH_m is referred to as an enthalpy change for the reaction. In this context the symbol Δ (delta) signifies change in" while H is the symbol for the quantity being changed, namely the enthalpy. We will deal with the enthalpy in some detail in Chap. 15. For the moment we can think of it as a property of matter which increases when matter absorbs energy and decreases when matter releases energy.

It is important to realize that the value of ΔH_m given in thermochemical above equations depends on the physical state of both the reactants and the products. Thus, if water were obtained as a gas instead of a liquid in the reaction in above equation, the value of ΔH_m would be different from -890.4 kJ. It is also necessary to specify both the temperature and pressure since the value of ΔH_m depends very slightly on these variables. If these are not specified they usually refer to 25°C and to normal atmospheric pressure.

Two more characteristics of thermochemical equations arise from the law of conservation of energy. The first is that *writing an equation in the reverse direction changes the sign of the enthalpy change*. For example,

$$H_2O(l) \rightarrow H_2O(g)$$
$$\Delta H_m = 44\,kJ$$

In the image above, the flames input energy into the water, giving it the energy necessary to transition to the gas phase. Since flames provide the energy for the phase transition, this is an endothermic reaction (energy is absorbed).

Tells us that when a mole of liquid water vaporizes, 44 kJ of heat is absorbed. This corresponds to the fact that heat is absorbed from your skin when perspiration evaporates, and you cool off. Condensation of 1 mol of water vapor, on the other hand, gives off exactly the same quantity of heat.

$$H_2O(g) \rightarrow H_2O(l)$$
$$\Delta H_m = -44kJ$$

It's counterintuitive, but the common summer occurrence seen above is actually exothermic. Since the reaction isn't highly exothermic (like the combustion of CH_4), we find it hard to associate with a release of energy. Thermodynamics allows us to better understand on a micro level energy changes like this one.

To see why this must be true, suppose that ΔH_m first equation = 44 kJ while ΔH_m second equation = −50.0 kJ. If we took 1 mol of liquid water and allowed it to evaporate, 44 kJ would be absorbed. We could then condense the water vapor, and 50.0 kJ would be given off. We could again have 1 mol of liquid water at 25°C but we would also have 6 kJ of heat which had been created from nowhere! This would violate the law of conservation of energy. The only way the problem can he avoided is for ΔH_m of the reverse reaction to be equal in magnitude but opposite in sign from ΔH_m of the forward reaction. That is,

$$\Delta H_m forward = -\Delta H_m reverse$$

Exergonic and Endergonic Reaction

Endergonic reactions are processes in physical chemistry or thermochemistry. This type of reaction needs an input of energy to create the products, which have more free energy as a whole than the sum of free energy of each of the reactants. An endothermic reaction is an endergonic reaction that involves heat or thermal energy in the process.

Endergonic Reactions

Endergonic reactions are nonspontaneous, as they need an input of energy to occur. An example of this type of reaction in biology is photosynthesis. This process depends on the reaction to absorb energy in the form of sunshine from the surroundings for it to occur. Plants capture some of the sun's energy as sunlight and use it to produce glucose from water and carbon dioxide. Glucose has more free energy than the reactants of carbon dioxide and water. The chemical bonds formed in an endergonic reaction are weaker than the chemical bonds that are broken. For this reason, it may also be called an unfavorable reaction, as it requires more energy input than you get from the end product. Another example of an endergonic reaction occurs when ice as a solid is melted by heat into liquid water, which is also called endothermic because the results are driven by warmer temperatures.

Exergonic Reactions

An exergonic reaction is called a spontaneous or a favorable reaction, and it is the opposite of an endergonic reaction. This type of reaction releases energy into the surroundings and forms stronger chemical bonds in the process than those that were broken in the reactants to produce the product. The free energy of the system decreases in an exergonic reaction. Some examples include mixing chlorine and sodium to make ordinary table salt and chemiluminescence when visible light is the energy that is released in the process. When the temperature of the surroundings rises, the reaction is exothermic as well as exergonic.

Endergonic and Exergonic Reactions

An endergonic and endothermic reaction occurs when energy is absorbed from the surroundings. In endothermic reactions, heat is absorbed. If you mix sodium carbonate (baking soda) and citric acid in water, the liquid becomes cold, but not cold enough to cause frostbite.

An exergonic reaction releases energy to the surroundings, and when it does, it is in the form of heat—it is exothermic. An example of this can be seen when you do laundry. Put a small amount of laundry detergent in your hand and add a small amount of water to it. You feel warmth emitting from the mixture, as it is an exothermic and exergonic reaction.

An example of an exergonic reaction that is not exothermic is a glow stick. Instead of releasing heat into the surroundings, it emits light.

Equilibrium Constant

The equilibrium constant for the reaction is related to ΔG^o by the relation:

$$K = e^{-\frac{\Delta G^\circ}{RT}}$$

where T is the absolute temperature and R is the gas constant. A positive value of ΔG^o therefore implies

$$K < 1$$

so that starting from molar stoichiometric quantities such a reaction would move backwards toward equilibrium, not forwards.

Nevertheless, endergonic reactions are quite common in nature, especially in biochemistry and physiology. Examples of endergonic reactions in cells include protein synthesis, and the Na^+/K^+ pump which drives nerve conduction and muscle contraction.

Gibbs Free Energy for Endergonic Reactions

All physical and chemical systems in the universe follow the second law of thermodynamics and proceed in a downhill, i.e., *exergonic*, direction. Thus, left to itself, any physical or chemical system will proceed, according to the second law of thermodynamics, in a direction that tends to lower the free energy of the system, and thus to expend energy in the form of work. These reactions occur spontaneously.

A chemical reaction is endergonic when non spontaneous. Thus in this type of reactions the Gibbs free energy increases. The entropy is included in any change of the Gibbs free energy. This differs from an endothermic reaction where the entropy is not included. The Gibbs free energy is calculated with the Gibbs–Helmholtz equation:

$$\Delta G = \Delta H - T \cdot \Delta S$$

where:

T = temperature in kelvins (K)

ΔG = change in the Gibbs free energy

ΔS = change in entropy (at 298 K) as $\Delta S = \Sigma\{S(\text{Product})\} - \Sigma\{S(\text{Reagent})\}$

ΔH = change in enthalpy (at 298 K) as $\Delta H = \Sigma\{\text{H(Product)}\} - \Sigma\{\text{H(Reagent)}\}$

A chemical reaction progresses non spontaneously when the Gibbs free energy increases, in that case the ΔG is positive. In exergonic reactions the ΔG is negative and in endergonic reactions the ΔG is positive:

$$\Delta_R G < 0 \text{ exergon}$$

$$\Delta_R G > 0 \text{ endergon}$$

where:

$\Delta_R G$ equals the change in the Gibbs free energy after completion of a chemical reaction.

Making Endergonic Reactions Happen

Endergonic reactions can be achieved if they are either *pulled* or *pushed* by an exergonic (stability increasing, negative change in free energy) process. Of course, in all cases the net reaction of the *total* system (the reaction under study plus the puller or pusher reaction) is exergonic.

Pull

Reagents can be *pulled* through an endergonic reaction, if the reaction products are cleared rapidly by a subsequent exergonic reaction. The concentration of the products of the endergonic reaction thus always remains low, so the reaction can proceed.

A classic example of this might be the first stage of a reaction which proceeds via a transition state. The process of getting to the top of the activation energy barrier to the transition state is endergonic. However, the reaction can proceed because having reached the transition state, it rapidly evolves via an exergonic process to the more stable final products.

Push

Endergonic reactions can be *pushed* by coupling them to another reaction which is strongly exergonic, through a shared intermediate.

This is often how biological reactions proceed. For example, on its own the reaction

$$X + Y \rightarrow XY$$

may be too endergonic to occur. However it may be possible to make it occur by coupling it to a strongly exergonic reaction – such as, very often, the decomposition of ATP into ADP and inorganic phosphate ions, ATP $\rightarrow$ ADP + P_i, so that

$$X + ATP \rightarrow XP + ADP$$
$$XP + Y \rightarrow XY + P_i$$

This kind of reaction, with the ATP decomposition supplying the free energy needed to make an endergonic reaction occur, is so common in cell biochemistry that ATP is often called the "universal energy currency" of all living organisms.

Isodesmic Reaction

In isodesmic reactions, the number and types of bonds are conserved on the reactant and product sides of the reaction. This again ensures a very similar amount of electron correlation on both sides of the reaction and leads to more accurate thermochemical values from even low-level computational methods.

An example isodesmic reaction is $CH_2F_2 + CH_4 \rightarrow 2\ CH_3F$. On the reactant side, there are two C-F single bonds and six C-H single bonds. On the product side, there are the same number of C-F and C-H bonds as on the reactant side.

Another example of an isodesmic reaction is $NH_3 + CH_3NH_3^+ \rightarrow NH_4^+ + CH_3NH_2$. A reaction of this type could be used to determine the proton affinity of nitrogen bases relative to ammonia.

To illustrate the use of isodesmic reactions, consider a comparison of ring strain in cyclopropane and ethylene oxide.

To explore this computationally, we might design an isodesmic reaction like the one shown below.

In this reaction, the three-membered ring on the reactant side is cyclopropane and on the product side it is ethylene oxide. If the reaction is calculated to be exothermic, that would suggest that ethylene oxide is more stable (less strained) than cyclopropane. If it is endothermic, then ethylene oxide is presumed to be less stable than cyclopropane. The energies for the noncyclic molecules are assumed to include no ring strain. Since the reactant and product sides both include two C-O bonds, three C-C bonds, and 12 C-H bonds, the energy difference is expected to result primarily from differences in ring strain. Results for this reaction, calculated at the Hartree-Fock/6-31G(d) level of theory, are presented in table.

Hartree-Fock/6-31G(d) enthalpies

Species	$E_e + H_{corr}$ **(kcal/mol)**
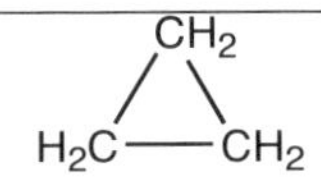	–73400.8
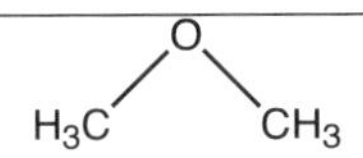	–96623.1
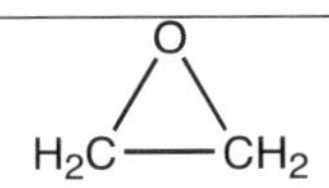	–95886.7
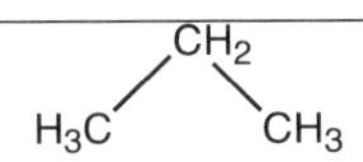	–74142.3
ΔH_{rx}	**–5.1**

The reaction enthalpy is calculated to be exothermic, –5.1 kcal/mol. This implies that the product side is more stable. In other words, this isodesmic reaction predicts that ethylene oxide has less ring strain than cyclopropane.

References

- Guibet, J.-C. Carburants et moteurs. Publication de l'Institut Français du Pétrole. ISBN 2-7108-0704-1
- Thermochemistry, physical-chemistry: tutorvista.com, Retrieved 29 May 2018
- What-is-an-endergonic-reaction-13712147: sciencing.com, Retrieved 19 July 2018
- Schmidt-Rohr, K (2015). "Why Combustions Are Always Exothermic, Yielding About 418 kJ per Mole of O2". J. Chem. Educ. 92: 2094–2099. Bibcode:2015JChEd..92.2094S. doi:10.1021/acs.jchemed.5b00333
- Chemical-thermodynamics-definition-principles: study.com, Retrieved 17 May 2018

3

Equilibrium Chemistry

Equilibrium chemistry studies systems that are in chemical equilibrium. In order to completely understand equilibrium chemistry, it is essential to understand the concepts central to it, such as chemical equilibrium, equilibrium constant and the law of chemical equilibrium, which have been discussed in detail in this chapter.

Systems in chemical equilibrium are characterized by minimum value of the free energy, which when plotted against the reaction coordinate has a slope that is zero. This is true for all chemical mixtures in equilibrium and may be applied to acid-base, metal-complex, host-guest, partition, solubility, chromatography and redox equilibria. An important consequence of this principle is the measurement of the equilibrium constant of a reaction, which can provide vital information about the composition of a system at equilibrium. It also aids in the understanding of acid-base homeostasis and biochemical processes of the human body.

Chemical Equilibrium

Chemical equilibrium, a condition in the course of a reversible chemical reaction in which no net change in the amounts of reactants and products occurs. A reversible chemical reaction is one in which the products, as soon as they are formed, react to produce the original reactants. At equilibrium, the two opposing reactions go on at equal rates, or velocities, hence there is no net change in the amounts of substances involved. At this point the reaction may be considered to be completed; *i.e.*, for some specified reaction condition, the maximum conversion of reactants to products has been attained.

The conditions that pertain to equilibrium may be given quantitative formulation. For example, for the reversible reaction $A \leftrightarrows B + C$, the velocity of the reaction to the right, r_1, is given by the mathematical expression (based on the law of mass action) $r_1 = k_1(A)$, where k_1 is the reaction-rate constant and the symbol in parentheses represents the concentration of A. The velocity of the reaction to the left, r_2, is $r_2 = k_2(B)(C)$. At equilibrium, $r_1 = r_2$, therefore:

$$k_1(A)_e = k_2(B)_e(C)_e \text{ or}$$
$$k_1 / k_2 = (B)_e(C)_e / (A)_e = K.$$

The subscript *e* represents conditions at equilibrium. For a given reaction, at some specified condition of temperature and pressure, the ratio of the amounts of products and reactants present at equilibrium, each raised to their respective powers, is a constant, designated the equilibrium constant of the reaction and represented by the symbol *K*. The value of the equilibrium constant varies with the temperature and pressure according to the principle of Le Chatelier.

By methods of statistical mechanics and chemical thermodynamics, it can be shown that the equilibrium constant is related to the change in the thermodynamic quantity called the standard Gibbs free energy accompanying the reaction. The standard Gibbs free energy of the reaction, ΔG^o, which is the difference between the sum of the standard free energies of the products and that of the reactants, is equal to the negative natural logarithm of the equilibrium constant multiplied by the so-called gas constant *R* and the absolute temperature *T*:

$$\Delta G^o = -RT \text{ In } K.$$

The equation allows the calculation of the equilibrium constant, or the relative amounts of products and reactants present at equilibrium, from measured or derived values of standard free energies of substances.

Thermodynamics

At constant temperature and pressure, one must consider the Gibbs free energy, *G*, while at constant temperature and volume, one must consider the Helmholtz free energy: *A*, for the reaction; and at constant internal energy and volume, one must consider the entropy for the reaction: *S*.

The constant volume case is important in geochemistry and atmospheric chemistry where pressure variations are significant. Note that, if reactants and products were in standard state (completely pure), then there would be no reversibility and no equilibrium. Indeed, they would necessarily occupy disjoint volumes of space. The mixing of the products and reactants contributes a large entropy (known as entropy of mixing) to states containing equal mixture of products and reactants. The standard Gibbs energy change, together with the Gibbs energy of mixing, determine the equilibrium state.

Here the constant pressure case is considered. The relation between the Gibbs free energy and the equilibrium constant can be found by considering chemical potentials.

At constant temperature and pressure, the Gibbs free energy, *G*, for the reaction depends only on the extent of reaction: ξ (Greek letter xi), and can only decrease according to the second law of thermodynamics. It means that the derivative of *G* with ξ must be negative if the reaction happens; at the equilibrium the derivative being equal to zero.

$$\left(\frac{dG}{d\xi}\right)_{T,p} = 0: \quad \text{equilibrium}$$

In order to meet the thermodynamic condition for equilibrium, the Gibbs energy must be stationary, meaning that the derivative of G with respect to the extent of reaction: ξ, must be zero. It can be shown that in this case, the sum of chemical potentials of the products is equal to the sum of those corresponding to the reactants. Therefore, the sum of the Gibbs energies of the reactants must be the equal to the sum of the Gibbs energies of the products.

$$\alpha\mu_{\mathrm{A}} + \beta\mu_{\mathrm{B}} = \sigma\mu_{\mathrm{S}} + \tau\mu_{\mathrm{T}}$$

where μ is in this case a partial molar Gibbs energy, a chemical potential. The chemical potential of a reagent A is a function of the activity, {A} of that reagent.

$$\mu_{\mathrm{A}} = \mu_{A}^{\ominus} + RT\,ln\{\mathrm{A}\}$$

(where $\mu^{\mathrm{o}}{}_{\mathrm{A}}$ is the standard chemical potential).

The definition of the Gibbs energy equation interacts with the fundamental thermodynamic relation to produce,

$$dG = Vdp - SdT + \sum_{i=1}^{k} \mu_i dN_i.$$

Inserting $dN_i = \nu_i\, d\xi$ into the above equation gives a Stoichiometric coefficient (υ_{i}) and a differential that denotes the reaction occurring once ($d\xi$). At constant pressure and temperature the above equations can be written as,

$$\left(\frac{dG}{d\xi}\right)_{T,p} = \sum_{i=1}^{k} \mu_i \nu_i = \Delta_{\mathrm{r}} G_{T,p}$$ which is the "Gibbs free energy change for the reaction.

This results in:

$$\Delta_{\mathrm{r}} G_{T,p} = \sigma\mu_{\mathrm{S}} + \tau\mu_{\mathrm{T}} - \alpha\mu_{\mathrm{A}} - \beta\mu_{\mathrm{B}}.$$

By substituting the chemical potentials:

$$\Delta_{\mathrm{r}} G_{T,p} = (\sigma\mu_{\mathrm{S}}^{\ominus} + \tau\mu_{\mathrm{T}}^{\ominus}) - (\alpha\mu_{\mathrm{A}}^{\ominus} + \beta\mu_{\mathrm{B}}^{\ominus}) + (\sigma RT\,ln\{\mathrm{S}\} + \tau RT\,ln\{\mathrm{T}\}) - (\alpha RT\,ln\{\mathrm{A}\} + \beta RT\,ln\{\mathrm{B}\}$$

the relationship becomes:

$$\Delta_{\mathrm{r}} G_{T,p} = \sum_{i=1}^{k} \mu_i^{\ominus} \nu_i + RT\,ln \frac{\{\mathrm{S}\}^{\sigma}\{\mathrm{T}\}^{\tau}}{\{\mathrm{A}\}^{\alpha}\{\mathrm{B}\}^{\beta}}$$

$$\sum_{i=1}^{k} \mu_i^{\ominus} \nu_i = \Delta_{\mathrm{r}} G^{\ominus}$$

which is the standard Gibbs energy change for the reaction that can be calculated using thermodynamical tables. The reaction quotient is defined as:

$$Q_{\mathrm{r}} = \frac{\{S\}^{\sigma}\{T\}^{\tau}}{\{A\}^{\alpha}\{B\}^{\beta}}$$

Therefore,

$$\left(\frac{dG}{d\xi}\right)_{T,p} = \Delta_{\mathrm{r}} G_{T,p} = \Delta_{\mathrm{r}} G^{\ominus} + RT\,ln\,Q_{\mathrm{r}}$$

At equilibrium:

$$\left(\frac{dG}{d\xi}\right)_{T,p} = \Delta_{\mathrm{r}} G_{T,p} = 0$$

leading to:

$$0 = \Delta_{\mathrm{r}} G^{\ominus} + RT\,ln\,K_{\mathrm{eq}}$$

and,

$$\Delta_{\mathrm{r}} G^{\ominus} = -RT\,ln\,K_{\mathrm{eq}}$$

Obtaining the value of the standard Gibbs energy change, allows the calculation of the equilibrium constant.

Addition of Reactants or Products

- For a reactional system at equilibrium: $Q_{\mathrm{r}} = K_{\mathrm{eq}}$; $\xi = \xi_{\mathrm{eq}}$.
- If are modified activities of constituents, the value of the reaction quotient changes and becomes different from the equilibrium constant: $Q_{\mathrm{r}} \neq K_{\mathrm{eq}}$

$$\left(\frac{dG}{d\xi}\right)_{T,p} = \Delta_{\mathrm{r}} G^{\ominus} + RT\,ln\,Q_{\mathrm{r}}$$

and,

$$\Delta_{\mathrm{r}} G^{\ominus} = -RT\,ln\,K_{eq}$$

then,

$$\left(\frac{dG}{d\xi}\right)_{T,p} = RT\,ln\left(\frac{Q_{\mathrm{r}}}{K_{\mathrm{eq}}}\right)$$

- If activity of a reagent *i* increases,

  $$Q_r = \frac{\prod (a_j)^{v_j}}{\prod (a_i)^{v_i}},$$ the reaction quotient decreases.

  then,

  $$Q_r < K_{eq} \quad \text{and} \quad \left(\frac{dG}{d\xi}\right)_{T,p} < 0$$

  The reaction will shift to the right (i.e. in the forward direction, and thus more products will form).

- If activity of a product *j* increases

  then,

  $$Q_r > K_{eq} \quad \text{and} \quad \left(\frac{dG}{d\xi}\right)_{T,p} > 0$$

The reaction will shift to the left (i.e. in the reverse direction, and thus less products will form).

Note that activities and equilibrium constants are dimensionless numbers.

Treatment of Activity

The expression for the equilibrium constant can be rewritten as the product of a concentration quotient, K_c and an activity coefficient quotient, Γ.

$$K = \frac{[S]^\sigma [T]^\tau \ldots}{[A]^\alpha [B]^\beta \ldots} \times \frac{\gamma_S{}^\sigma \gamma_T{}^\tau \ldots}{\gamma_A{}^\alpha \gamma_B{}^\beta \ldots} = K_c \Gamma$$

[A] is the concentration of reagent A, etc. It is possible in principle to obtain values of the activity coefficients, γ. For solutions, equations such as the Debye–Hückel equation or extensions such as Davies equation Specific ion interaction theory or Pitzer equations may be used. However this is not always possible. It is common practice to assume that Γ is a constant, and to use the concentration quotient in place of the thermodynamic equilibrium constant. It is also general practice to use the term *equilibrium constant* instead of the more accurate *concentration quotient*. This practice will be followed here.

For reactions in the gas phase partial pressure is used in place of concentration and fugacity coefficient in place of activity coefficient. In the real world, for example, when making ammonia in industry, fugacity coefficients must be taken into account. Fugacity, *f*, is the product of partial pressure and fugacity coefficient. The chemical potential of a species in the gas phase is given by,

$$\mu = \mu^{\ominus} + RT\,ln\left(\frac{f}{\text{bar}}\right) = \mu^{\ominus} + RT\,ln\left(\frac{p}{\text{bar}}\right) + RT\,ln\,\gamma$$

so the general expression defining an equilibrium constant is valid for both solution and gas phases.

Concentration Quotients

In aqueous solution, equilibrium constants are usually determined in the presence of an "inert" electrolyte such as sodium nitrate $NaNO_3$ or potassium perchlorate $KClO_4$. The ionic strength of a solution is given by,

$$I = \frac{1}{2}\sum_{i=1}^{N} c_i z_i^2$$

where c_i and z_i stand for the concentration and ionic charge of ion type i, and the sum is taken over all the N types of charged species in solution. When the concentration of dissolved salt is much higher than the analytical concentrations of the reagents, the ions originating from the dissolved salt determine the ionic strength, and the ionic strength is effectively constant. Since activity coefficients depend on ionic strength the activity coefficients of the species are effectively independent of concentration. Thus, the assumption that Γ is constant is justified. The concentration quotient is a simple multiple of the equilibrium constant.

$$K_c = \frac{K}{\Gamma}$$

However, K_c will vary with ionic strength. If it is measured at a series of different ionic strengths the value can be extrapolated to zero ionic strength. The concentration quotient obtained in this manner is known, paradoxically, as a thermodynamic equilibrium constant.

To use a published value of an equilibrium constant in conditions of ionic strength different from the conditions used in its determination, the value should be adjusted.

Metastable Mixtures

A mixture may appear to have no tendency to change, though it is not at equilibrium. For example, a mixture of SO_2 and O_2 is metastable as there is a kinetic barrier to formation of the product, SO_3.

$$2\,SO_2 + O_2 \rightleftharpoons 2\,SO_3$$

The barrier can be overcome when a catalyst is also present in the mixture as in the contact process, but the catalyst does not affect the equilibrium concentrations.

Likewise, the formation of bicarbonate from carbon dioxide and water is very slow under normal conditions,

$$CO_2 + 2H_2O \rightleftharpoons HCO_3^- + H_3O^+$$

but almost instantaneous in the presence of the catalytic enzyme carbonic anhydrase.

Pure Substances

When pure substances (liquids or solids) are involved in equilibria their activities do not appear in the equilibrium constant because their numerical values are considered one.

Applying the general formula for an equilibrium constant to the specific case of a dilute solution of acetic acid in water one obtains,

$$\mathrm{CH_3CO_2H + H_2O \rightleftharpoons CH_3CO_2^- + H_3O^+}$$

$$K_c = \frac{[\mathrm{CH_3CO_2^-}][\mathrm{H_3O^+}]}{[\mathrm{CH_3CO_2H}][\mathrm{H_2O}]}$$

For all but very concentrated solutions, the water can be considered a "pure" liquid, and therefore it has an activity of one. The equilibrium constant expression is therefore usually written as,

$$K = \frac{[\mathrm{CH_3CO_2^-}][\mathrm{H_3O^+}]}{[\mathrm{CH_3CO_2H}]} = K_c.$$

A particular case is the self-ionization of water itself,

$$\mathrm{2\,H_2O \rightleftharpoons H_3O^+ + OH^-}$$

Because water is the solvent, and has an activity of one, the self-ionization constant of water is defined as,

$$K_w = [\mathrm{H^+}][\mathrm{OH^-}]$$

It is perfectly legitimate to write $[H^+]$ for the hydronium ion concentration, since the state of solvation of the proton is constant (in dilute solutions) and so does not affect the equilibrium concentrations. K_w varies with variation in ionic strength and/or temperature.

The concentrations of H^+ and OH^- are not independent quantities. Most commonly $[OH^-]$ is replaced by $K_w[H^+]^{-1}$ in equilibrium constant expressions which would otherwise include hydroxide ion.

Solids also do not appear in the equilibrium constant expression, if they are considered to be pure and thus their activities taken to be one. An example is the Boudouard reaction:

$$2\,CO \rightleftharpoons CO_2 + C$$

for which the equation (without solid carbon) is written as:

$$K_c = \frac{[CO_2]}{[CO]^2}$$

Multiple Equilibria

Consider the case of a dibasic acid H_2A. When dissolved in water, the mixture will contain H_2A, HA^- and A^{2-}. This equilibrium can be split into two steps in each of which one proton is liberated.

$$H_2A \rightleftharpoons HA^- + H^+: \quad K_1 = \frac{[HA-][H+]}{[H2A]}$$

$$HA^- \rightleftharpoons A^{2-} + H^+: \quad K_2 = \frac{[A^{2-}][H+]}{[HA-]}$$

K_1 and K_2 are examples of *stepwise* equilibrium constants. The *overall* equilibrium constant, β_D, is product of the stepwise constants.

$$H_2A \rightleftharpoons A^{2-} + 2H^+: \quad \beta_D = \frac{[A^{2-}][H^+]^2}{[H_2A]} = K_1K_2$$

Note that these constants are dissociation constants because the products on the right hand side of the equilibrium expression are dissociation products. In many systems, it is preferable to use association constants.

$$A^{2-} + H+ \rightleftharpoons HA^-: \quad \beta_1 = \frac{[HA^-]}{[A^{2-}][H+]}$$

$$A^{2-} + 2H+ \rightleftharpoons H_2A: \quad \beta_2 = \frac{[H2A]}{[A^{2-}][H+]^2}$$

β_1 and β_2 are examples of association constants. Clearly $\beta_1 = \frac{1}{K_2}$ and $\beta_2 = \frac{1}{\beta_D}$; $\log\beta_1 = pK_2$ and $\log\beta_2 = pK_2 + pK_1$.

Effect of Temperature

The effect of changing temperature on an equilibrium constant is given by the van 't Hoff equation,

$$\frac{d\ln K}{dT} = \frac{\Delta H_m^{\ominus}}{RT^2}$$

Thus, for exothermic reactions (ΔH is negative), K decreases with an increase in temperature, but, for endothermic reactions, (ΔH is positive) K increases with an increase temperature. An alternative formulation is,

$$\frac{d \ln K}{d(T^{-1})} = -\frac{\Delta H^{\ominus}_{\mathrm{m}}}{R}$$

At first sight this appears to offer a means of obtaining the standard molar enthalpy of the reaction by studying the variation of K with temperature. In practice, however, the method is unreliable because error propagation almost always gives very large errors on the values calculated in this way.

Effect of Electric and Magnetic Fields

The effect of electric field on equilibrium has been studied by Manfred Eigen among others.

Types of Equilibrium

Haber–Bosch process

1. N_2 (g) $\rightleftharpoons$ N_2 (adsorbed)
2. N_2 (adsorbed) $\rightleftharpoons$ 2 N (adsorbed)
3. H_2 (g) $\rightleftharpoons$ H_2 (adsorbed)
4. H_2 (adsorbed) $\rightleftharpoons$ 2 H (adsorbed)
5. N (adsorbed) + 3 H(adsorbed) $\rightleftharpoons$ NH_3 (adsorbed)
6. NH_3 (adsorbed) $\rightleftharpoons$ NH_3 (g)

- In the gas phase: rocket engines
- The industrial synthesis such as ammonia in the Haber–Bosch process (depicted right) takes place through a succession of equilibrium steps including adsorption processes
- Atmospheric chemistry
- Seawater and other natural waters: chemical oceanography
- Distribution between two phases
  - log D distribution coefficient: important for pharmaceuticals where lipophilicity is a significant property of a drug
  - Liquid–liquid extraction, Ion exchange, Chromatography
  - Solubility product

- Uptake and release of oxygen by hemoglobin in blood
- Acid–base equilibria: acid dissociation constant, hydrolysis, buffer solutions, indicators, acid–base homeostasis
- Metal–ligand complexation: sequestering agents, chelation therapy, MRI contrast reagents, Schlenk equilibrium
- Adduct formation: host–guest chemistry, supramolecular chemistry, molecular recognition, dinitrogen tetroxide
- In certain oscillating reactions, the approach to equilibrium is not asymptotically but in the form of a damped oscillation
- The related Nernst equation in electrochemistry gives the difference in electrode potential as a function of redox concentrations
- When molecules on each side of the equilibrium are able to further react irreversibly in secondary reactions, the final product ratio is determined according to the Curtin–Hammett principle.

In these applications, terms such as stability constant, formation constant, binding constant, affinity constant, association/dissociation constant are used. In biochemistry, it is common to give units for binding constants, which serve to define the concentration units used when the constant's value was determined.

Composition of a Mixture

When the only equilibrium is that of the formation of a 1:1 adduct as the composition of a mixture, there are many ways that the composition of a mixture can be calculated.

There are three approaches to the general calculation of the composition of a mixture at equilibrium.

- The most basic approach is to manipulate the various equilibrium constants until the desired concentrations are expressed in terms of measured equilibrium constants (equivalent to measuring chemical potentials) and initial conditions.
- Minimize the Gibbs energy of the system.
- Satisfy the equation of mass balance. The equations of mass balance are simply statements that demonstrate that the total concentration of each reactant must be constant by the law of conservation of mass.

Mass-balance Equations

In general, the calculations are rather complicated or complex. For instance, in the case of a dibasic acid, H_2A dissolved in water the two reactants can be specified as the conjugate base, A^{2-}, and the proton, H^+. The following equations of mass-balance could apply equally well to a base such as 1,2-diaminoethane, in which case the base itself is designated as the reactant A:

$$T_A = [A] + [HA] + [H_2A]$$
$$T_H = [H] + [HA] + 2[H_2A] - [OH]$$

With T_A the total concentration of species A. Note that it is customary to omit the ionic charges when writing and using these equations.

When the equilibrium constants are known and the total concentrations are specified there are two equations in two unknown "free concentrations" [A] and [H]. This follows from the fact that $[HA] = \beta_1[A][H]$, $[H_2A] = \beta_2[A][H]^2$ and $[OH] = K_w[H]^{-1}$,

$$T_A = [A] + \beta_1[A][H] + \beta_2[A][H]^2$$
$$T_H = [H] + \beta_1[A][H] + 2\beta_2[A][H]^2 - K_w[H]^{-1}$$

so the concentrations of the "complexes" are calculated from the free concentrations and the equilibrium constants. General expressions applicable to all systems with two reagents, A and B would be,

$$T_A = [A] + \sum_i p_i \beta_i [A]^{p_i} [B]^{q_i}$$
$$T_B = [B] + \sum_i q_i \beta_i [A]^{p_i} [B]^{q_i}$$

It is easy to see how this can be extended to three or more reagents.

Polybasic Acids

Species concentrations during hydrolysis of the aluminium.

The composition of solutions containing reactants A and H is easy to calculate as a function of p[H]. When [H] is known, the free concentration [A] is calculated from the mass-balance equation in A.

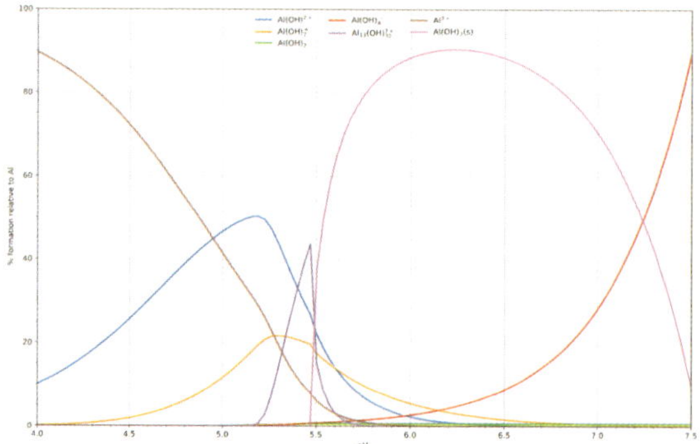

The diagram alongside, shows an example of the hydrolysis of the aluminium Lewis acid $Al^{3+}_{(aq)}$ shows the species concentrations for a 5×10^{-6} M solution of an aluminium salt as a function of pH. Each concentration is shown as a percentage of the total aluminium.

Solution and Precipitation

The diagram above illustrates the point that a precipitate that is not one of the main species in the solution equilibrium may be formed. At pH just below 5.5 the main species present in a 5 μM solution of Al^{3+} are aluminium hydroxides $Al(OH)^{2+}$, $AlOH^{+}_{2}$ and $Al_{13}(OH)^{7+}_{32}$, but on raising the pH $Al(OH)_3$ precipitates from the solution. This occurs because $Al(OH)_3$ has a very large lattice energy. As the pH rises more and more $Al(OH)_3$ comes out of solution. This is an example of Le Châtelier's principle in action: Increasing the concentration of the hydroxide ion causes more aluminium hydroxide to precipitate, which removes hydroxide from the solution. When the hydroxide concentration becomes sufficiently high the soluble aluminate, $Al(OH)^{-}_{4}$, is formed.

Another common instance where precipitation occurs is when a metal cation interacts with an anionic ligand to form an electrically neutral complex. If the complex is hydrophobic, it will precipitate out of water. This occurs with the nickel ion Ni^{2+} and dimethylglyoxime, ($dmgH_2$): in this case the lattice energy of the solid is not particularly large, but it greatly exceeds the energy of solvation of the molecule $Ni(dmgH)_2$.

Minimization of Gibbs Energy

At equilibrium, at a specified temperature and pressure, the Gibbs energy G is at a minimum:

$$dG = \sum_{j=1}^{m} \mu_j \, dN_j = 0$$

For a closed system, no particles may enter or leave, although they may combine in various ways. The total number of atoms of each element will remain constant. This means that the minimization above must be subjected to the constraints:

$$\sum_{j=1}^{m} a_{ij} N_j = b_i^{o}$$

where a_{ij} is the number of atoms of element i in molecule j and b^{o}_{i} is the total number of atoms of element i, which is a constant, since the system is closed. If there are a total of k types of atoms in the system, then there will be k such equations. If ions are involved, an additional row is added to the a_{ij} matrix specifying the respective charge on each molecule which will sum to zero.

This is a standard problem in optimisation, known as constrained minimisation. The most common method of solving it is using the method of Lagrange multipliers, also known as undetermined multipliers (though other methods may be used).

$$\mathcal{G} = G + \sum_{i=1}^{k} \lambda_i \left(\sum_{j=1}^{m} a_{ij} N_j - b_i^{o} \right) = 0$$

where the λ_i are the Lagrange multipliers, one for each element. This allows each of the N_j and λ_j to be treated independently, and it can be shown using the tools of multivariate calculus that the equilibrium condition is given by,

$$0 = \frac{\partial \mathcal{G}}{\partial N_j} = \mu_j + \sum_{i=1}^{k} \lambda_i a_{ij}$$

$$0 = \frac{\partial \mathcal{G}}{\partial \lambda_i} = \sum_{j=1}^{m} a_{ij} N_j - b_i^{o}$$

This is a set of (m + k) equations in (m + k) unknowns (the N_j and the λ_i) and may, therefore, be solved for the equilibrium concentrations N_j as long as the chemical potentials are known as functions of the concentrations at the given temperature and pressure. Note that the second equation is just the initial constraints for minimization.

This method of calculating equilibrium chemical concentrations is useful for systems with a large number of different molecules. The use of k atomic element conservation equations for the mass constraint is straightforward, and replaces the use of the stoichiometric coefficient equations. The results are consistent with those specified by chemical equations. For example, if equilibrium is specified by a single chemical equation:

$$\sum_{j=0}^{m} \nu_j R_j = 0$$

where ν_j is the stochiometric coefficient for the j th molecule (negative for reactants, positive for products) and R_j is the symbol for the j th molecule, a properly balanced equation will obey:

$$\sum_{j=1}^{m} a_{ij} \nu_j = 0$$

Multiplying the first equilibrium condition by ν_j yields,

$$0 = \sum_{j=1}^{m} \nu_j \mu_j + \sum_{j=1}^{m} \sum_{i=1}^{k} \nu_j \lambda_i a_{ij} = \sum_{j=1}^{m} \nu_j \mu_j$$

As above, defining ΔG,

$$\Delta G=\sum_{j=1}^{m}\nu_j\mu_j=\sum_{j=1}^{m}\nu_j(\mu_j^{\ominus}+RT\,ln(\{R_j\}))=\Delta G^{\ominus}+RT\,ln\left(\prod_{j=1}^{m}\{R_j\}^{\nu j}\right)=\Delta G^{\ominus}+RT\,ln(Q_r)$$

which will be zero at equilibrium.

Homogeneous Equilibria

A homogeneous equilibrium has the whole thing present in the similar phase. The common examples contain reactions where everything is a gas, or everything is present in the similar solution. Examples of few homogeneous equilibrium systems are given below:

(1) Dissociation of nitrogen tetroxide: The dissociation of nitrogen tetroxide is represented by the equation,

$N_2O_4\ (g) \leftrightarrow 2NO_2$

for which the equilibrium constant is,

$K_p = [(P_{NO2})^2 / P_{N2O4}]$

P_{NO2} and P_{N2O4} being the partial pressure of NO_2 and N_2O_4 respectively.

The experimental method applied in determining the equilibrium constant consists in measuring the total pressure at equilibrium in a flask of known volume in which a known mass of N_2O_4 was placed. Using the ideal gas equation, the apparent molecular mass of the partially dissociated gas is calculated and equation is used to determine the degree of dissociation at equilibrium. If a represents the degree of dissociation, (1- α) is proportional to the number of moles of undissociated N_2O_4 and 2α is proportional to the number of moles of NO_2 so that (1- α + 2α) or (1 + a) is proportional to the total number of moles.

If P is the total pressure, the partial pressures are given by,

$P_{NO2} = \{2\alpha / (1 + \alpha)\}P$ and $P_{N2O4} = \{2\alpha / (1 + \alpha)\}P$

Hence, $K_p = [\{2\alpha / (1 + \alpha)\}P]^2 / \{2\alpha / (1 + \alpha)\}P$

$= \{4\alpha^2 / (1 - \alpha^2)\}P$

(2) Dissociation of phosphorus pentachloride:

Phosphorous pentachloride dissociates on heating:

$PCI_5\,(g) \Leftrightarrow PCI_3\,(g) + CI_2\,(g)$

If $K_{c\,=}$ 3.26 x 10^{-2} at 191°C, what is K_p at this temperature?

Ans.: K_p = 1.24

When heated, phosphorus pentachloride dissociates and the reaction is represented by the following equation:

$$PCl_5\ (g) \leftrightarrow PCl_3\ (g) + Cl_2\ (g)$$

The equilibrium constant for this reaction is,

$$K_C = [PCl_3]\ [Cl_2]\ /\ [PC1_5]$$

which can he determined by measuring the degree of dissociation al the given temperature as in the case of the dissociation of nitrogen tetroxide.

(3) Formation and decomposition of hydrogen iodide gas:

A classical example of gaseous homogeneous equilibrium is the following reaction above 400°C.

$$2\ HI\ (g) \leftrightarrow H_2\ (g) + I_2\ (g)$$

This reaction was extensively studied by Bodenstein and others. Known quantities of hydrogen iodide were sealed in quartz vessels of known volume and heated at the desired temperature for sufficient time until equilibrium was reached. The vessels were then suddenly cooled and the contents analyzed for iodine by titration with standard sodium thiosulphate solution. As the equation above shows, the concentration of hydrogen is the same as that of iodine while the concentration of hydrogen iodide at equilibrium is the initial concentration of hydrogen iodide minus twice the concentration of iodine.

Heterogeneous Equilibria

When the state of equilibrium in a system has components in more than one phase it is termed as a heterogeneous equilibrium. For example, if we take a container with ice and water at a temperature that is allowing the existence of both the phases simultaneously, such that, both ice and water are present in a state of equilibrium. This state is termed as heterogeneous equilibrium.in terms of its equation,

In terms of its equation, this can be written as:

$$H_2O(s) \leftrightarrow H_2O(l)$$

In another example, we can consider an aqueous solution of a solid such as calcium hydroxide. We notice that the solid calcium hydroxide is in equilibrium with its saturated solution.

Writing the equilibrium constant for heterogeneous reactions is different from that of the homogeneous reactions. For example, consider the thermal dissociation of calcium carbonate into calcium oxide and carbon dioxide. Here, we can see that the equilibrium constant for the dissociation of the reactant into its products is only dependent on the gaseous component, as the solid and the liquid reactants are considered to be constant.

$$CaCO_3(s) \leftrightarrow CaO(s) + CO_2$$

Here, the components $CaCO_3$ and CaO are solids, so their molar concentration remains constant throughout the reaction. Therefore the equilibrium constant can be written as,

$$K_c = \frac{[CaO][CO_2]}{[CaCO_3]}$$

$$K_c = [CO_2]$$

Also, in terms of K_p, we can write

$$K_p = pCO_2$$

Where p denotes the partial pressure. In other words, we can state that, at a given temperature, there is a constant concentration or partial pressure of CO_2 in the equilibrium reaction with CaO and $CaCO_3$.

Example: What's the concentration of Ca^{+2} ions in a saturated solution of $CaCO_3$? K_{sp} ($CaCO_3$) = 4.5×10^{-9}.

Solution: In the equation for the dissociation of $CaCO_3$, we see that the concentration of Ca^{+2} and CO_3^{2-} must be the same in a saturated solution because one of each ion is formed by the breakup of each $CaCO_3$. Because we don't know what this concentration will be, let's call it "x."

Putting this into our K_{sp} expression for $CaCO_3$, we find that:

$$K_{sp} = 4.5 \times 10^{-9} = x^2$$

$$x = 6.7 \times 10^{-5} \text{ M.}$$

The concentration of both the calcium and carbonate ions is 6.7×10^{-5} in a saturated $CaCO_3$ solution.

Equilibrium Constant

Consider the reversible chemical reaction:

$$aA + bB \leftrightharpoons cC + dD$$

A and B are the reactants, C and D are the products and a, b, c, and d are the coefficients from the balanced reaction.

When the rate of the forward reaction equals the rate of the reverse reaction, the system is in chemical equilibrium. It is useful to know how much of each substance is in the container - in particular the amount of products compared to the amount of reactants. A simple ratio from the balanced chemical equation gives us a number called the equilibrium constant (K_c).

$$K_c = \frac{[C]^c\,[D]^d}{[A]^a\,[B]^b}$$

[A], [B], [C] and [D] [A], [B], [C] and [D] are the molar concentrations of each substance present at equilibrium.

When the concentration of the reactants is much larger than the concentration of the products K_c will be small (normally less than one). When the concentration of the reactants is much less than that of the products K_c will be large (normally greater than one).

For example:

$$2NO(g)+2H_2\,(g)\rightleftharpoons N_2\,(g)+2H_2\,O(g)$$

The reactants are NO and H_2. The products are N_2 and H_2O.

It is important to look at the coefficients of the equation as well:

$$2NO(g) + 2H_2\,(g) \rightleftharpoons 1\,N_2\,(g) + 2\,H_2\,O(g)$$

K_c for this equation will therefore be written as follows:

$$K_c = \frac{[N_2]^1\,[H_2O]^2}{[NO]^2\,[H_2]^2}$$

In the expression for K_c the concentration of a product or reactant is taken to the power of its coefficient in the balanced reaction. So if the coefficient of a product (C) is 3 in the balanced equation, then the concentration of C ([C]) will be written $[C]^3$ in the expression for K_c.

We leave out reactants or products that are either pure liquids or in the solid phase when calculating K_c. For example:

$$C(s)+H_2O(g) \rightleftharpoons CO(g)+H_2(g) K_c = \frac{[CO][H_2]}{[H_2O]}$$

Calculating the Equilibrium Constant (ESCNK)

When calculations involving the equilibrium constant are done, the following tips may help:

- Always *read the question carefully* to be sure that you understand what you have been asked to calculate.
- If the equilibrium constant is involved, make sure that the concentrations you use are the concentrations at equilibrium, and not the concentrations or quantities that are present at some other time in the reaction.
- When you are doing more complicated calculations, it helps to draw up a RICE table.

Remember that the general form of the expression for K_c of a reaction at chemical equilibrium ($aA + bB \leftrightharpoons cC + dD$) is:

$$K_c = \frac{[C]^c[D]^d}{[A]^a[B]^b}$$

Worked example 1: Writing expressions for K_c

Question

For the reaction:

$$9X(g) + Y\ 3(g) \rightleftharpoons 3X_3Y(g)$$

Write an expression for the equilibrium constant K_c.

Solution

Step 1: Write the general expression for K_c

$$K_c = \frac{[C]^c[D]^d}{[A]^a[B]^b}$$

Step 2: Determine the reactants and the products of the reaction

X(g) and Y_3 (g) are both reactants. They are gases and will be included in the expression.

X3 Y(g) is a product. It is a gas and will be included in the expression.

Step 3: Write the general expression for K_c for this reaction

$$K_c = \frac{[X_3Y]^z}{[X]^x[Y_3]^y}$$

Step 4: What are the coefficients in the balanced equation?

The compound X has a coefficient of 9. The compound Y3 has a coefficient of 1. The compound X3Y has a coefficient of 3.

Step 5: Write the expression for K_c for this reaction

$$K_c = \frac{[X_3Y]^3}{[X]^9[Y_3]^1}$$

Worked example 2: Calculating reagent concentration

Question

For the reaction:

$$S(s)+O_2(g) \leftrightharpoons SO_2\,(g)$$

Write an expression for the equilibrium constant.

Calculate the equilibrium concentration of O_2 if: $K_c = 6$ and $[SO_2] = 3$ mol.dm^{-3}at equilibrium.

Solution

Step 1: Determine which compounds will be part of the K_c expression

$S(s) + O_2\,(g) \leftrightharpoons SO_2(g)$

O_2 and SO_2 are gases and so will be part of the expression for K_c. S is solid and so will not be part of the expression for K_c.

Step 2: Write the expression for K_c

$$K_c = \frac{[SO_2]}{[O_2]}$$

Step 3: Re-arrange the expression so that oxygen is on its own on one side of the expression

$$[O_2] = \frac{[SO_2]}{K_c}$$

Step 4: Fill in the values you know and calculate $[O_2\,]$

$$[O_2] = \frac{3\text{mol.dm}^{-3}}{6} = 0{,}5\text{ mol.dm}^{-3}$$

Worked example 3: Calculating K_c

Question

For the reaction:

$$SO_2\,(g) + NO_2\,(g) \rightleftharpoons NO(g) + SO_3(g)$$

The concentrations of the compounds used are:

$[SO_2]$ = 0, 2 mol.dm^{-3}, $[NO_2]$ = 0,1 mol.dm^{-3}, [NO] =0,4 mol.dm^{-3} and

$[SO_3]$ = 0, 2 mol.dm^{-3}

Calculate the value of K_c.

Solution

Step 1: Write the general expression for K_c

$$K_c = \frac{[c]^c [D]^d}{[A]^a [B]^b}$$

Step 2: Determine the reactants and the products of the reaction

SO_2 (g) and NO_2 (g) are both reactants. They are gases and will be included in the expression.

NO(g) and SO_3 (g) are both products. They are gases and will be included in the expression.

Step 3: Write the expression for K_c for this reaction

All four compounds have a coefficient of 1.

$$K_c = \frac{[NO]^1 [SO]^1}{[SO_2]^1 [NO_2]^1}$$

Step 4: Fill in the values you know for this expression and calculate K_c

$$K_c = \frac{(0,4)(0,2)}{(0,2)(0,1)} = 4$$

Types of Equilibrium Constants

Cumulative and Stepwise Formation Constants

A cumulative or overall constant, given the symbol β, is the constant for the formation of a complex from reagents. For example, the cumulative constant for the formation of ML_2 is given by,

$$M + 2\,L \rightleftharpoons ML_2; \qquad [ML_2] = \beta_{12}[M][L]^2$$

The stepwise constant, *K*, for the formation of the same complex from ML and L is given by,

$$ML + L \rightleftharpoons ML_2; \qquad [ML_2] = K[ML][L] = K\beta_{11}[M][L]^2$$

It follows that,

$$\beta_{12} = K\beta_{11}$$

A cumulative constant can always be expressed as the product of stepwise constants. There is no agreed notation for stepwise constants, though a symbol such as K^{L}_{ML} is sometimes found in the literature. It is best always to define each stability constant by reference to an equilibrium expression.

Competition Method

A particular use of a stepwise constant is in the determination of stability constant values outside the normal range for a given method. For example, EDTA complexes of many metals are outside the range for the potentiometric method. The stability constants for those complexes were determined by competition with a weaker ligand.

$$ML + L' \rightleftharpoons ML' + L \; [ML^{¢}]=K\frac{[ML][L']}{[L]}=K\frac{\beta_{ML}[M][L][L']}{[L]}=K\beta_{ML}[M][L']; \;\; \beta_{ML¢}=K\beta_{ML}$$

The formation constant of $[Pd(CN)_4]^{2-}$ was determined by the competition method.

Association and Dissociation Constants

In organic chemistry and biochemistry it is customary to use pK_a values for acid dissociation equilibria.

$$pK_a = -log\,K_{diss} = log\left(\frac{1}{K_{diss}}\right)$$

where *log* denotes a logarithm to base 10 or common logarithm, and K_{diss} is a stepwise acid dissociation constant. For bases, the base association constant, pK_b is used. For any given acid or base the two constants are related by pK_a + pK_b = pK_w, so pK_a can always be used in calculations.

On the other hand, stability constants for metal complexes, and binding constants for host–guest complexes are generally expressed as association constants. When considering equilibria such as,

$$M + HL \rightleftharpoons ML + H$$

it is customary to use association constants for both ML and HL. Also, in generalized computer programs dealing with equilibrium constants it is general practice to use cumulative constants rather than stepwise constants and to omit ionic charges from equilibrium expressions. For example, if NTA, nitrilotriacetic acid, $N(CH_2CO_2H)_3$ is designated as H_3L and forms complexes ML and MHL with a metal ion M, the following expressions would apply for the dissociation constants.

$$H_3L \rightleftharpoons H_2L+H; \quad pK_1=-\log\left(\frac{[H2L][H]}{[H3L]}\right)$$

$$H_2L \rightleftharpoons HL+H; \quad pK_2=-\log\left(\frac{[HL][H]}{[H2L]}\right)$$

$$HL \rightleftharpoons L+H; \quad pK_3=-\log\left(\frac{[L][H]}{[HL]}\right)$$

The cumulative association constants can be expressed as,

$$L+H \rightleftharpoons HL; \quad \log b_{011} = \log\left(\frac{[HL]}{[L][H]}\right) = pK_3$$

$$L+2H \rightleftharpoons H_2L; \quad \log b_{012} = \log\left(\frac{[H_2L]}{[L][H]^2}\right) = pK_3 + pK_2$$

$$L+3H \rightleftharpoons H_3L; \quad \log b_{013} = \log\left(\frac{[H_3L]}{[L][H]^3}\right) = pK_3 + pK_2 + pK_1$$

$$M+L \rightleftharpoons ML; \quad \log b_{110} = \log\left(\frac{[ML]}{[M][L]}\right)$$

$$M+L+H \rightleftharpoons MLH; \quad \log b_{111} = \log\left(\frac{[MLH]}{[M][L][H]}\right)$$

Note how the subscripts define the stoichiometry of the equilibrium product.

Micro-constants

When two or more sites in an asymmetrical molecule may be involved in an equilibrium reaction there are more than one possible equilibrium constants. For example, the molecule L-DOPA has two non-equivalent hydroxyl groups which may be deprotonated. Denoting L-DOPA as LH_2, the following diagram shows all the species that may be formed ($X = CH_2CH(NH_2)CO_2H$):

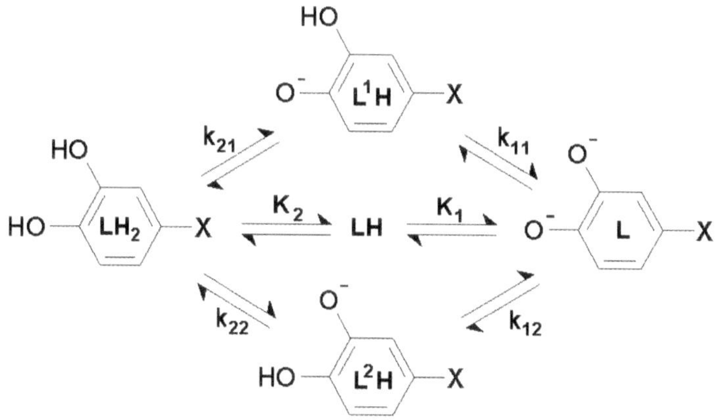

The first protonation constants are,

$$[L^1H] = k_{11}[L][H],\ [L^2H] = k_{12}[L][H]$$

The concentration of LH^- is the sum of the concentrations of the two micro-species. Therefore, the equilibrium constant for the reaction, the macro-constant, is the sum of the micro-constants.

$$K_1 = k_{11} + k_{12}$$

In the same way,

$$K_2 = k_{21} + k_{22}$$

Lastly, the cumulative constant is,

$$\beta_2 = K_1K_2 = k_{11}k_{21} = k_{12}k_{22}$$

Thus, although there are seven micro-and macro-constants, only three of them are mutually independent. Moreover, the isomerization constant, K_i, is equal to the ratio of the microconstants.

$$K_i = k_{11}/k_{12}$$

In L-DOPA the isomerization constant is 0.9, so the micro-species L^1H and L^2H have almost equal concentrations at all pH values.

In general a macro-constant is equal to the sum of all the micro-constants and the occupancy of each site is proportional to the micro-constant. The site of protonation can be very important, for example, for biological activity.

Micro-constants cannot be determined individually by the usual methods, which give macro-constants. Methods which have been used to determine micro-constants include:

- Blocking one of the sites, for example by methylation of a hydroxyl group, to determine one of the micro-constants.

- Using a spectroscopic technique, such as infrared spectroscopy, where the different micro-species give different signals.
- Applying mathematical procedures to ^{13}C NMR data.

pH Considerations (Brønsted Constants)

pH is defined in terms of the activity of the hydrogen ion,

$$pH = -\log_{10} \{H^+\}$$

In the approximation of ideal behaviour, activity is replaced by concentration. pH is measured by means of a glass electrode, a mixed equilibrium constant, also known as a Brønsted constant, may result.

$$HL \rightleftharpoons L + H; \qquad pK = -log\left(\frac{[L]\{H\}}{[HL]}\right)$$

It all depends on whether the electrode is calibrated by reference to solutions of known activity or known concentration. In the latter case the equilibrium constant would be a concentration quotient. If the electrode is calibrated in terms of known hydrogen ion concentrations it would be better to write p[H] rather than pH, but this suggestion is not generally adopted.

Hydrolysis Constants

In aqueous solution the concentration of the hydroxide ion is related to the concentration of the hydrogen ion by,

$$K_W=[H][OH]$$
$$[OH]=K_W[H]^{-1}$$

The first step in metal ion hydrolysis can be expressed in two different ways,

$$M(H2O)<=>M(OH)+H; \quad [M(OH)]=\beta^*[M][H]^{-1}$$
$$M+OH<=>M(OH); \quad [M(OH)]=K[M][OH]=KK_W[M][H]^{-1}$$

It follows that $\beta^* = KK_W$. Hydrolysis constants are usually reported in the β^* form and therefore often have values much less than 1. For example, if log K = 4 and log K_W = −14, log β^* = 4 + (−14) = −10 so that $\beta^* = 10^{-10}$. In general when the hydrolysis product contains n hydroxide groups log β^* = log K + n log K_W.

Conditional Constants

Conditional constants, also known as apparent constants, are concentration quotients which are not true equilibrium constants but can be derived from them. A very common instance is where pH is fixed at a particular value. For example, in the case of iron(III) interacting with EDTA, a conditional constant could be defined by,

$$K_{cond} = \frac{[\text{Total Fe bound to EDTA}]}{[\text{Total Fe not bound to EDTA}] \times [\text{Total EDTA not bound to Fe}]}$$

This conditional constant will vary with pH. It has a maximum at a certain pH. That is the pH where the ligand sequesters the metal most effectively.

In biochemistry equilibrium constants are often measured at a pH fixed by means of a buffer solution. Such constants are, by definition, conditional and different values may be obtained when using different buffers.

Gas-phase Equilibria

For equilibria in a gas phase, fugacity, f, is used in place of activity. However, fugacity has the dimension of pressure, so it must be divided by a standard pressure, usually 1 bar, in order to produce a dimensionless quantity, $\frac{f}{p^{e}}$. An equilibrium constant is expressed in terms of the dimensionless quantity. For example, for the equilibrium $2NO_2 \rightleftharpoons N_2O_4$,

$$\frac{f_{N_2O_4}}{p^{\ominus}} = K\left(\frac{f_{NO_2}}{p^{\ominus}}\right)^2$$

Fugacity is related to partial pressure, p, by a dimensionless fugacity coefficient ϕ: $f = \phi p$. Thus, for the example,

$$K = \frac{\phi_{N_2O_4} p_{N_2O_4} / p^{\ominus}}{\left(\phi_{NO_2} p_{NO_2} / p^{\ominus}\right)^2}$$

Usually the standard pressure is omitted from such expressions. Expressions for equilibrium constants in the gas phase then resemble the expression for solution equilibria with fugacity coefficient in place of activity coefficient and partial pressure in place of concentration.

$$K = \frac{\phi_{N_2O_4} p_{N_2O_4}}{\left(\phi_{NO_2} p_{NO_2}\right)^2}$$

Thermodynamic Basis for Equilibrium Constant Expressions

Thermodynamic equilibrium is characterized by the free energy for the whole (closed) system being a minimum. For systems at constant temperature and pressure the Gibbs free energy is minimum. The slope of the reaction free energy with respect to the reaction coordinate, ξ, is zero when the free energy is at its minimum value.

$$\left(\frac{\partial G}{\partial \xi}\right)_{T,P} = 0$$

The free energy change, δG_r, can be expressed as a weighted sum of chemical potentials, the partial molar free energies of the species in equilibrium. The potential, μ_i, of the *i*th species in a chemical reaction is the partial derivative of the free energy with respect to the number of moles of that species, N_i,

$$\mu_i = \left(\frac{\partial G}{\partial N_i}\right)_{T,P}$$

A general chemical equilibrium can be written as,

$$\sum_j n_j \text{Reactant}_j \rightleftharpoons \sum_k m_k \text{Product}_k$$

where n_j are the stoichiometric coefficients of the reactants in the equilibrium equation, and m_j are the coefficients of the products. At equilibrium,

$$\sum_k m_k \mu_k = \sum_j n_j \mu_j$$

The chemical potential, μ_i, of the *i*th species can be calculated in terms of its activity, a_i.

$$\mu_i = \mu_i^{\ominus} + RT \ln a_i$$

μ^o_i is the standard chemical potential of the species, *R* is the gas constant and *T* is the temperature. Setting the sum for the reactants *j* to be equal to the sum for the products, *k*, so that $\delta G_r(\text{Eq}) = 0$,

$$\sum_j n_j (\mu_j^{\ominus} + RT \ln a_j) = \sum_k m_k (\mu_k^{\ominus} + RT \ln a_k)$$

Rearranging the terms,

$$\sum_k m_k \mu_k^{\ominus} - \sum_j n_j \mu_j^{\ominus} = -RT \left(\sum_k \ln a_k{}^{mk} - \sum_j \ln a_j{}^{nj} \right)$$

$$\Delta G^{\ominus} = -RT \ln K.$$

This relates the standard Gibbs free energy change, ΔG° to an equilibrium constant, *K*, the reaction quotient of activity values at equilibrium.

$$\Delta G^{\ominus} = \sum_k m_k \mu_k^{\ominus} - \sum_j n_j \mu_j^{\ominus}$$

$$\ln K = \sum_k \ln a_k^{m_k} - \sum_j \ln a_j^{n_j} ; K = \frac{\prod_k a_k^{m_k}}{\prod_j a_j^{n_j}} \equiv \frac{\{R\}^{\rho}\{S\}^{\sigma}...}{\{A\}^{\alpha}\{B\}^{\beta}...}$$

Equivalence of Thermodynamic and Kinetic Expressions for Equilibrium Constants

At equilibrium the rate of the forward reaction is equal to the backward reaction rate. A simple reaction, such as ester hydrolysis,

$$AB + H2O \rightleftharpoons AH + B(OH)$$

has reaction rates given by expressions,

$$\text{forward rate} = k_f[AB][H2O]$$
$$\text{backward rate} = k_b[AH][B(OH)]$$

According to Guldberg and Waage, equilibrium is attained when the forward and backward reaction rates are equal to each other. In these circumstances, an equilibrium constant is defined to be equal to the ratio of the forward and backward reaction rate constants.

$$K = \frac{k_f}{k_b} = \frac{[AH][B(OH)]}{[AB][H_2O]}$$

For convenience the concentration of water may be taken to be constant, resulting in the simpler expression,

$$K^c = \frac{[AH][B(OH)]}{[AB]}$$

This particular concentration quotient, K^c, has the dimension of concentration, but the thermodynamic equilibrium constant, K, is always dimensionless.

Unknown Activity Coefficient Values

It is very rare for activity coefficient values to have been determined experimentally for a system at equilibrium. There are three options for dealing with the situation where activity coefficient values are not known from experimental measurements.

Use calculated activity coefficients, together with concentrations of reactants. For equilibria in solution estimates of the activity coefficients of charged species can be obtained using Debye–Hückel theory, an extended version, or SIT theory. For uncharged species, the activity coefficient γ_0 mostly follows a "salting-out" model: $\log_{10} \gamma_0 = bI$ where I stands for ionic strength.

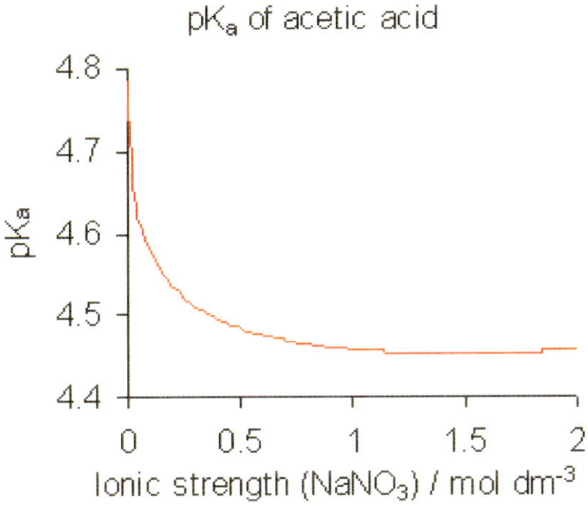

Variation of log K_c of acetic acid with ionic strength

Assume that the activity coefficients are all equal to 1. This is acceptable when all concentrations are very low.

For equilibria in solution use a medium of high ionic strength. In effect this redefines the standard state as referring to the medium. Activity coefficients in the standard state are, by definition, equal to 1. The value of an equilibrium constant determined in this manner is dependent on the ionic strength. When published constants refer to an ionic strength other than the one required for a particular application, they may be adjusted by means of specific ion theory (SIT) and other theories.

Dimensionality

Any equilibrium constant, K, must be dimensionless. An equilibrium constant is related to the standard Gibbs free energy change for the reaction by the expression,

$$\Delta G^{\ominus} = -RT \ln K$$

This requires that K be a pure number from which a logarithm can be derived. In the case of a simple equilibrium,

$$\mathrm{A+B} \rightleftharpoons \mathrm{AB}$$

the thermodynamic equilibrium constant is defined in terms of the activities, {AB}, {A} and {B}, of the species in equilibrium with each other.

$$K = \frac{\{\mathrm{AB}\}}{\{\mathrm{A}\}\{\mathrm{B}\}}$$

Now, each activity term can be expressed as a product of a concentration [X] and a corresponding activity coefficient, $\gamma(X)$. Therefore,

$$K = \frac{[AB]}{[A][B]} \times \frac{\gamma(AB)}{\gamma(A)\gamma(B)}$$

When each activity coefficient has the dimension 1/concentration, K remains dimensionless. However, when the quotient of activity coefficients is set equal to 1, K appears to have the dimension of 1/concentration.

$$K = \frac{[AB]}{[A][B]}$$

This is what usually happens in practice when an equilibrium constant is calculated as a quotient of concentration values. The assumption underlying this practice is that the quotient of activities is constant under the conditions in which the equilibrium constant value is determined. There conditions are usually achieved by keeping the reaction temperature constant and by using a medium of relatively high ionic strength as the solvent. It is not unusual, particularly in texts relating to biochemical equilibria, to see a value quoted with a dimension. The justification for this practice is that the concentration scale used may be either mol dm^{-3} or mmol dm^{-3}, so that the concentration unit has to be stated in order to avoid there being any ambiguity.

In general equilibria between two reagents can be expressed as,

$$pA + qB \rightleftharpoons A_pB_q$$

in which case the equilibrium constant is defined, in terms of numerical concentration values, as,

$$K = \frac{[A_pB_q]}{[A]^p[B]^q}$$

The apparent dimension of this K value is concentration^{1-p-q}; this may be written as $M^{(1-p-q)}$ or $mM^{(1-p-q)}$, where the symbol M signifies a molar concentration (1 M = 1 mol dm^{-3})

A further complications arises from the fact that an equilibrium constant can be defined as an association constant or as a dissociation constant. For dissociation constants,

$$A_pB_q <=> pA + qB; K = \frac{[A]^p[B]^q}{[A_pB_q]}$$

The apparent dimension associated with a dissociation constant is the reciprocal of the dimension associated with the corresponding association constant.

Concentration Scales

When concentrations are expressed on the scale of mole fractions, all terms in the expression,

$$K=\frac{[AB]}{[A][B]}\times\frac{\gamma(AB)}{\gamma(A)\gamma(B)}=\frac{[AB]}{[A][B]}\times\Gamma$$

are pure numbers, so there is no issue regarding dimensionality. However, it is the usual practice to measure concentrations in solution on the scale of mol dm^{-3}. In that case, the quotient $\frac{[AB]}{[A][B]}$ appears to have the dimension of dm^3 mol^{-1}. It is therefore necessary to take dimensionality into account in discussing the thermodynamics of chemical equilibria. There are two possible approaches.

Set the dimension of Γ to be the reciprocal of the dimension of the concentration quotient. This is almost universal practice in the field of stability constant determinations. The "equilibrium constant" $\frac{K}{\Gamma}$, is dimensionless. It will be a function of the ionic strength of the medium used for the determination. However, this process is inconsistent with Debye–Hückel theory, where $\gamma(AB)$, *etc.* are taken to be pure numbers. The numerical value of Γ may be designated as 1. This is equivalent to re-defining the standard states. With this convention $\frac{K}{\Gamma}\equiv K$.

Replace each concentration term $[X]$ by the dimensionless quotient $\frac{[X]}{[X^{o}]}$, where $[X^{o}]$ is the concentration of reagent *X* in its standard state. By definition the numerical value of $\tilde{a}(X^{o})$ is 1, so Γ also has a numerical value of 1.

In both approaches the numerical value of the stability constant is unchanged. The first is more useful for practical purposes; in fact, the unit of the concentration quotient is often attached to a published stability constant value in the biochemical literature. The second approach is consistent with the standard exposition of Debye–Hückel theory.

Water as Both Reactant and Solvent

For reactions in aqueous solution, such as an acid dissociation reaction,

$$AH + H_2O \rightleftharpoons A^- + H_3O^+$$

the concentration of water may be taken as being constant and the formation of the hydronium ion is implicit.

$$AH \rightleftharpoons A^- + H^+$$

Water concentration is omitted from expressions defining equilibrium constants, except when solutions are very concentrated.

$K = \frac{[A][H]}{[AH]}$ (K defined as a dissociation constant).

Similar considerations apply to metal ion hydrolysis reactions.

Enthalpy and Entropy: Temperature Dependence

If both the equilibrium constant, K and the standard enthalpy change, $\Delta H^{\ominus}$, for a reaction have been determined experimentally, the standard entropy change for the reaction is easily derived. Since $\Delta G = \Delta H - T\Delta S$ and $\Delta G = -RT \ln K$,

$$\Delta S^{\ominus} = \frac{\Delta H^{\ominus} + RT \ln K}{T}$$

To a first approximation the standard enthalpy change is independent of temperature. Using this approximation, definite integration of the van 't Hoff equation,

$$\Delta H^{\ominus} = -R \frac{d \ln K}{d(1/T)}$$

gives,

$$\ln K_2 = \ln K_1 - \frac{\Delta H^{\ominus}}{R}\left(\frac{1}{T_2} - \frac{1}{T_1}\right)$$

This equation can be used to calculate the value of log K at a temperature, T_2, knowing the value at temperature T_1.

The van 't Hoff equation also shows that, for an exothermic reaction ($\Delta H < 0$), when temperature increases K decreases and when temperature decreases K increases, in accordance with Le Chatelier's principle. The reverse applies when the reaction is endothermic.

When K has been determined at more than two temperatures, a straight line fitting procedure may be applied to a plot of $\ln K$ against $1/T$ to obtain a value for $\Delta H^{\ominus}$. Error propagation theory can be used to show that, with this procedure, the error on the calculated $\Delta H^{\ominus}$ value is much greater than the error on individual log K values. Consequently, K needs to be determined to high precision when using this method.For example, with a silver ion-selective electrode each log K value was determined with a precision of ca. 0.001 and the method was applied successfully.

Standard thermodynamic arguments can be used to show that, more generally, enthalpy will change with temperature.

$$\left(\frac{\partial H}{\partial T}\right)_p = C_p$$

where C_p is the heat capacity at constant pressure.

A More Complex Formulation

The calculation of K at a particular temperature from a known K at another given temperature can be approached as follows if standard thermodynamic properties are available. The effect of temperature on equilibrium constant is equivalent to the effect of temperature on Gibbs energy because:

$$\ln K = \frac{-\Delta_r G^{\ominus}}{RT}$$

where $\Delta_r G^{\ominus}$ is the reaction standard Gibbs energy, which is the sum of the standard Gibbs energies of the reaction products minus the sum of standard Gibbs energies of reactants.

Here, the term "standard" denotes the ideal behaviour (i.e., an infinite dilution) and a hypothetical standard concentration (typically 1 mol/kg). It does not imply any particular temperature or pressure because, although contrary to IUPAC recommendation, it is more convenient when describing aqueous systems over wide temperature and pressure ranges.

The standard Gibbs energy (for each species or for the entire reaction) can be represented (from the basic definitions) as:

$$G_{T_2}^{\ominus} = G_{T_1}^{\ominus} - S_{T_1}^{\ominus}(T_2 - T_1) - T_2 \int_{T_1}^{T_2} \frac{C_p^{\ominus}}{T} dT + \int_{T_1}^{T_2} C_p^{\ominus} dT$$

In the above equation, the effect of temperature on Gibbs energy (and thus on the equilibrium constant) is ascribed entirely to heat capacity. To evaluate the integrals in this equation, the form of the dependence of heat capacity on temperature needs to be known.

If the standard molar heat capacity $C_p^{\ominus}$ can be approximated by some analytic function of temperature (e.g. the Shomate equation), then the integrals involved in calculating other parameters may be solved to yield analytic expressions for them. For example, using approximations of the following forms:

For pure substances (solids, gas, liquid):

$$C_p^{\ominus} \approx A + BT + CT^{-2}$$

For ionic species at $T < 200$ °C:

$$C_p^{\ominus} \approx (4.186a + b\breve{S}_{T_1}^{\ominus}) \frac{(T_2 - T_1)}{ln\left(\frac{T_2}{T_1}\right)}$$

then the integrals can be evaluated and the following final form is obtained:

$$G^{\ominus}_{T_2} \approx G^{\ominus}_{T_1} + (C^{\ominus}_{p} - S^{\ominus}_{T_1})(T_2 - T_1) - T_2\, ln\left(\frac{T_2}{T_1}\right) C^{\ominus}_{p}$$

The constants A, B, C, a, b and the absolute entropy, $S^{o}_{298\ K}$, required for evaluation of $C^{o}_{p}(T)$, as well as the values of $G_{298\ K}$ and $S_{298\ K}$ for many species are tabulated in the literature.

Pressure Dependence

The pressure dependence of the equilibrium constant is usually weak in the range of pressures normally encountered in industry, and therefore, it is usually neglected in practice. This is true for condensed reactant/products (i.e., when reactants and products are solids or liquid) as well as gaseous ones.

For a gaseous-reaction example, one may consider the well-studied reaction of hydrogen with nitrogen to produce ammonia:

$$N_2 + 3\ H_2 \rightleftharpoons 2\ NH_3$$

If the pressure is increased by an addition of an inert gas, then neither the composition at equilibrium nor the equilibrium constant are appreciably affected (because the partial pressures remain constant, assuming an ideal-gas behaviour of all gases involved). However, the composition at equilibrium will depend appreciably on pressure when:

- The pressure is changed by compression of the gaseous reacting system, and
- The reaction results in the change of the number of moles of gas in the system.

In the example reaction above, the number of moles changes from 4 to 2, and an increase of pressure by system compression will result in appreciably more ammonia in the equilibrium mixture. In the general case of a gaseous reaction:

$$\alpha\, A + \beta\, B \rightleftharpoons \sigma\, S + \tau\, T$$

the change of mixture composition with pressure can be quantified using:

$$K_p = \frac{p_S{}^{\sigma} p_T{}^{\tau}}{p_A{}^{\alpha} p_B{}^{\beta}} = \frac{X_S{}^{\sigma} X_T{}^{\tau}}{X_A{}^{\alpha} X_B{}^{\beta}} P^{\sigma+\tau-\alpha-\beta} = K_X P^{\sigma+\tau-\alpha-\beta}$$

where p denote the partial pressures of the components, P is the total system pressure, X denote the mole fraction, K_p is the equilibrium constant expressed in terms of partial pressures and K_X is the equilibrium constant expressed in terms of mole fractions.

The above change in composition is in accordance with Le Chatelier's principle and does not involve any change of the equilibrium constant with the total system pressure. Indeed, for ideal-gas reactions K_p is independent of pressure.

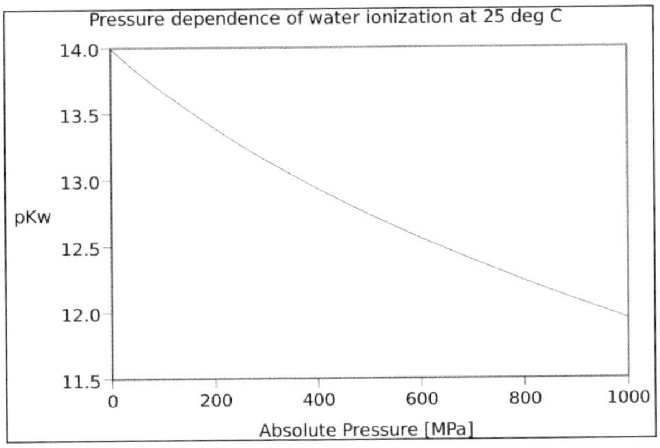

Pressure dependence of the water ionization constant at 25°C. In general, ionization in aqueous solutions tends to increase with increasing pressure.

In a condensed phase, the pressure dependence of the equilibrium constant is associated with the reaction molar volume. For reaction:

$$\alpha\, \mathrm{A} + \beta\, \mathrm{B} \rightleftharpoons \sigma\, \mathrm{S} + \tau\, \mathrm{T}$$

the reaction molar volume is:

$$\Delta\bar{V} = \sigma\bar{V}_{\mathrm{S}} + \tau\bar{V}_{\mathrm{T}} - \alpha\bar{V}_{\mathrm{A}} - \beta\bar{V}_{\mathrm{B}}$$

where $\bar{V}$ denotes a partial molar volume of a reactant or a product.

For the above reaction, one can expect the change of the reaction equilibrium constant (based either on mole-fraction or molal-concentration scale) with pressure at constant temperature to be:

$$\left(\frac{\partial \ln K_X}{\partial P}\right)_T = \frac{-\Delta\bar{V}}{RT}.$$

The matter is complicated as partial molar volume is itself dependent on pressure.

Effect of Isotopic Substitution

Isotopic substitution can lead to changes in the values of equilibrium constants, especially if hydrogen is replaced by deuterium (or tritium). This *equilibrium isotope effect* is analogous to the kinetic isotope effect on rate constants, and is primarily due to the change in zero-point vibrational energy of H–X bonds due to the change in mass upon isotopic substitution. The zero-point energy is inversely proportional to the square root of the mass of the vibrating hydrogen atom, and will therefore be smaller for a D–X bond that for an H–X bond.

An example is a hydrogen atom abstraction reaction R' + H–R ⇌ R'–H + R with equilibrium constant K_H, where R' and R are organic radicals such that R' forms a stronger bond to hydrogen than does R. The decrease in zero-point energy due to deuterium substitution will then be more important for R'–H than for R–H, and R'–D will be stabilized more than R–D, so that the equilibrium constant K_D for R' + D–R ⇌ R'–D + R is greater than K_H. This is summarized in the rule *the heavier atom favors the stronger bond.*

Similar effects occur in solution for acid dissociation constants (K_a) which describe the transfer of H^+ or D^+ from a weak aqueous acid to a solvent molecule: HA + H_2O = H_3O^+ + A^- or DA + D_2O ⇌ D_3O^+ + A^-. The deuterated acid is studied in heavy water, since if it were dissolved in ordinary water the deuterium would rapidly exchange with hydrogen in the solvent.

The product species H_3O^+ (or D_3O^+) is a stronger acid than the solute acid, so that it dissociates more easily and its H–O (or D–O) bond is weaker than the H–A (or D–A) bond of the solute acid. The decrease in zero-point energy due to isotopic substitution is therefore less important in D_3O^+ than in DA so that $K_D < K_H$, and the deuterated acid in D_2O is weaker than the non-deuterated acid in H_2O. In many cases the difference of logarithmic constants $pK_D - pK_H$ is about 0.6, so that the pD corresponding to 50% dissociation of the deuterated acid is about 0.6 units higher than the pH for 50% disssociation of the non-deuterated acid.

Law of Chemical Equilibrium

The concentration of reactants and products, at equilibrium, are constant at a given temperature. Do the equilibrium concentrations of these substances have any relation? Can equilibrium constant be expressed in other terms? Let's try to answer these questions.

Consider the following simple reversible reaction where A and B are reactants, C and D are the products.

$$A + B \rightleftharpoons C + D$$

A mixture of products and reactants at the equilibrium state is called equilibrium mixture. From experimental studies, it is observed that the concentration of chemical equilibrium mixtures has a relation. It can be represented by equilibrium equation and is given as follows:

$$K_c = \frac{[C][D]}{[A][B]}$$

K_c is called the equilibrium constant. The concentration of A at equilibrium is represented as [A] (similarly for B, C, and D), the stoichiometric coefficients of the reactants and products are 1. Now, what if the coefficients are different? Will the equation change? Well, it is experimentally observed that the equation does depend on the coefficients and is given by the law of chemical equilibrium.

The product of concentrations of the products raised to the respective stoichiometric coefficient in the balanced chemical equation divided by the product of concentrations of the reactants raised to their individual stoichiometric coefficients has a constant value (equilibrium constant) at a given temperature. This is known as the equilibrium law or law of chemical equilibrium or law of mass action.

For a balanced reaction of the type,

$$a\,A + b\,B \rightleftharpoons c\,C + d\,D$$

According to the law of mass action, the constant value obtained by relating equilibrium concentrations of reactants and products is called equilibrium constant. For the forward reaction, this is given by,

$$K_c = \frac{[C]^c [D]^d}{[A]^a [B]^b}$$

The equilibrium constant for the reverse reaction is the inverse of forward reaction and is given by:

$$K_c' = \frac{1}{K_c} = \frac{[C]^c [D]^d}{[A]^a [B]^b}$$

If the coefficients of the chemical equation are multiplied by a factor 'n' then the equilibrium constant is raised by the power 'n' i.e. the constant becomes K_c^n.

Equilibrium Constant Representation	**Expressed in terms of**	**Expressed as**
K_c	Concentrations of reactants and products	$\frac{[C]^c [D]^d}{[A]^a [B]^b}$
K_p	Partial pressures of reactants and products. (only for the substances which are in gaseous state)	$\frac{p_C^d p_D^d}{p_A^a p_B^b}$
K_x	Mole fractions of reactants and products	$\frac{[X_C]^c [X_D]^d}{[X_A]^a [X_B]^b}$

Relation between K_c, K_p and K_x

$$K_p = K_c \ (RT)^{\Delta n_g}$$

$$K_x = K_p \ (RT)^{\Delta n_g}$$

Where,

Δng = moles of gaseous products – moles of gaseous reactants.

K_c is the equilibrium constant expressed regarding concentration, similarly K_p is shown regarding partial pressures of the substances and K_x is expressed regarding mole fraction.

References

- Atkins, Peter; De Paula, Julio (2006). Atkins' Physical Chemistry (8th ed.). W. H. Freeman. pp. 200–202. ISBN 0-7167-8759-8
- Law-of-mass-action-or-law-of-chemical-equilibrium, chemistry: byjus.com, Retrieved 5 May 2018
- Clugston, Michael J. (1990). "A mathematical verification of the second law of thermodynamics from the entropy of mixing". Journal of Chemical Education. 67 (3): 203. Bibcode:-1990JChEd..67Q.203C. doi:10.1021/ed067p203
- chemistry, heterogeneous-equilibrium: byjus.com, Retrieved 15 July 2018
- "Project: Ionic Strength Corrections for Stability Constants". International Union of Pure and Applied Chemistry. Archived from the original on 29 October 2008. Retrieved 2008-11-23
- Homogeneous-equilibria, chemistry: qsstudy.com, Retrieved 19 May 2018
- Hague, David N.; Moreton, Anthony D. (1994). "Protonation sequence of linear aliphatic polyamines by 13C NMR spectroscopy". J. Chem. Soc., Perkin Trans. 2 (2): 265–70. doi:10.1039/P29940000265
- Chemical-equilibrium, science: britannica.com, Retrieved 9 July 2018
- Denbigh, K. (1981). "Chapter 4". The principles of chemical equilibrium (4th ed.). Cambridge: Cambridge University Press. ISBN 0-521-28150-4
- Schultz, Mary Jane (1999). "Why Equilibrium? Understanding Entropy of Mixing". Journal of Chemical Education. 76 (10): 1391. Bibcode:1999JChEd..76.1391S. doi:10.1021/ed076p1391

4

Gases: Theory and Laws

Gas is a fundamental state of matter. The properties of gases under different conditions of temperature and pressure are understood from ideal gas or real gas models. This chapter has been carefully written to provide an easy understanding of the different aspects of ideal and real gas, gas laws, kinetic theory of gases and the combined gas law.

Gas

A gas is a sample of matter that conforms to the shape of a container in which it is held and acquires a uniform density inside the container, even in the presence of gravity and regardless of the amount of substance in the container. If not confined to a container, gaseous matter, also known as vapor, will disperse into space . The term gas is also used in reference to the state, or condition, of matter having this property.

The atom s or molecule s of matter in the gaseous state move freely among each oth-er, and are, in most instances, packed more loosely than the molecules of the same substance in the solid or liquid state. A sample of gaseous matter can be compressed. Examples of gases are oxygen at room temperature (approximately 20°C or 68°F), hydrogen at room temperature, and water at standard atmospheric pressure and a temperature above 100°C or 212°F.

When a sample of matter in the gaseous state is heated, the atoms or molecules gain kinetic energy and move more rapidly. When a sample of gaseous matter is cooled, the atoms or molecules lose kinetic energy and move more slowly. If a sample of gaseous matter, confined to a container of fixed size, is heated, the pressure increases. If the sample is cooled, the pressure drops. If a sample of gaseous matter is placed in a sealed container and then the volume of the container is reduced, the compression heats the gas. If the volume of the sealed container increases, the decompression cools the gas.

If the temperature becomes sufficiently high, certain gases, such as hydrogen, will rapidly combine with other gases such as oxygen or chlorine. This is combustion. Some chemical reactions between gases and other substances occur more slowly; an example is the gradual oxidation of iron to form iron oxide (rust). In this case, the oxygen is gaseous at room temperature, while the iron and the iron oxide are solids.

When a sample of matter in the gaseous state is cooled to a low enough temperature, it may become a liquid or a solid. For example, if nitrogen is chilled to a temperature far below zero Celsius , it liquefies. Liquid nitrogen is used by some medical doctors to destroy minor skin lesions such as warts. Another gas, carbon dioxide, skips the liquid phase when cooled at atmospheric pressure, and becomes a solid known as dry ice.

Kinetic-molecular Picture

Gases nevertheless do have a structure of sorts on a molecular scale. They consist of a vast number of molecules moving chaotically in all directions and colliding with one another and with the walls of their container. Beyond this, there is no structure—the molecules are distributed essentially randomly in space, traveling in arbitrary directions at speeds that are distributed randomly about an average determined by the gas temperature. The pressure exerted by a gas is the result of the innumerable impacts of the molecules on the container walls and appears steady to human senses because so many collisions occur each second on all sections of the walls. More subtle properties such as heat conductivity, viscosity (resistance to flow), and diffusion are attributed to the molecules themselves carrying the mechanical quantities of energy, momentum, and mass, respectively. These are called transport properties, and the rate of transport is dominated by the collisions between molecules, which force their trajectories into tortuous shapes. The molecular collisions are in turn controlled by the forces between the molecules and are described by the laws of mechanics.

Thus, gases are treated as a large collection of tiny particles subject to the laws of physics. Their properties are attributed primarily to the motion of the molecules and can be explained by the kinetic theory of gases. It is not obvious that this should be the case, and for many years a static picture of gases was instead espoused, in which the pressure, for instance, was attributed to repulsive forces between essentially stationary particles pushing on the container walls. How the kinetic-molecular picture finally came to be universally accepted is a fascinating piece of scientific history and is discussed briefly below in the section Kinetic theory of gases. Any theory of gas behaviour based on this kinetic model must also be a statistical one because of the enormous numbers of particles involved. The kinetic theory of gases is now a classical part of statistical physics and is indeed a sort of miniature display case for many of the fundamental concepts and methods of science. Such important modern concepts as distribution functions, cross sections, microscopic reversibility, and time-reversal invariance have their historical roots in kinetic theory, as does the entire atomistic view of matter.

Molecular Sizes

Molecular sizes can be estimated from the foregoing information on the intermolecular separation, speed, mean free path, and collision rate of gas molecules. It would seem logical that large molecules should have a better chance of colliding than do small molecules. The collision frequency and mean free path must therefore be related to molecular size. To find this relationship, consider a single molecule in motion; during a time interval t it will sweep out a certain volume, hitting any other molecules present in this so-called collision volume. If molecules are located by their centres and each molecule has a diameter d, then the collision volume will be a long cylinder of cross-sectional area πd^2. The cylinder must be sufficiently long to include enough molecules so that good statistics on the number of collisions are obtained, but otherwise the length does not matter. If the molecule is observed for a time t, then the length of the collision cylinder will be $\bar{v}t$, where $\bar{v}$ is the average speed of the molecule, and the volume of the cylinder will be $(\pi d^2)(\bar{v}t)$, the product of its cross-sectional area and its length. Every molecule in the cylinder will be struck within time t, so the number of molecules in the collision cylinder will equal the number of collisions that occur in time t. Each collision will put a kink in the cylinder, but this will not affect the results as long as the number of collisions is not too large. If the gas is uniform, the number of molecules per volume will be consistent throughout the entire gas. Suppose that there are N molecules in volume V; then there will be $(N/V)(\pi d^2)(\bar{v}t)$ molecules in the collision volume; this is the number of collisions in time t. The mean free path is equal to the total length of the collision cylinder divided by the number of collisions that occur in it:

$$l = \frac{(\bar{v}t)}{(N/V)(\pi d^2)(\bar{v}t)} = \frac{1}{(N/V)(\pi d^2)}.$$

Since l has been shown to be roughly 2.0×10^{-5} cm, d could be calculated if N/V was known.

It is relatively easy to find $(N/V)d^3$, from which both d and N/V can be determined. Recall that the volume of one gram of steam is about 1,600 times larger than the volume of one gram of liquid water. In other words, there are roughly 1,600 N molecules in a volume V of liquid, and, if the molecules are just touching (i.e., the separation distance between their centres is one molecular diameter), the volume V of the liquid is 1,600 Nd^3. When this equation for volume is combined with the above expression for l, the following values are obtained: $d = \pi\ (2.0 \times 10^{-5})/1{,}600 = 3.9 \times 10^{-8}$ cm = 3.9 Å, and $N/V = 1/\pi d^2 l = 1.0 \times 10^{19}$ molecules per cubic centimetre. Thus, a typical molecule is exceedingly small, and there is an impressively large number of them in one cubic centimetre of gas.

Between collisions, a gas molecule travels a distance of about $l/d = (2.0 \times 10^{-5})/(3.9 \times 10^{-8}) = 500$ times its diameter. Since it was calculated above that the average separation between molecules is about 10 times the molecular diameter, the mean free path is approximately 50 times greater than the mean molecular separation. Accordingly, a typical molecule passes roughly 50 other molecules before it hits one.

Properties of Gases

Compressibility

An internal combustion engine provides a good example of the ease with which gases can be compressed. In a typical four-stroke engine, the piston is first pulled out of the cylinder to create a partial vacuum, which draws a mixture of gasoline vapor and air into the cylinder. The piston is then pushed into the cylinder, compressing the gasoline/air mixture to a fraction of its original volume.

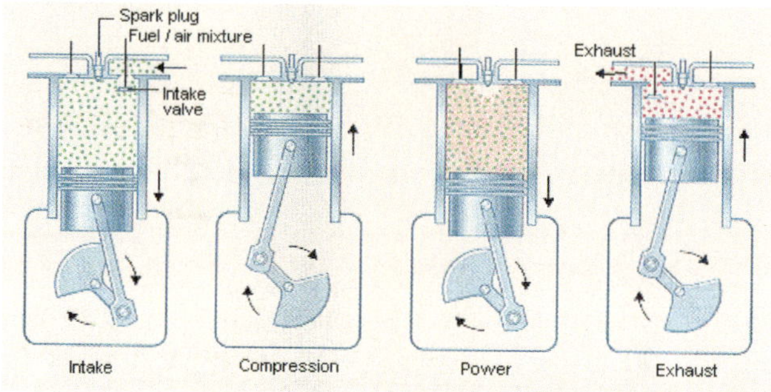

The operation of a four-stroke engine can be divided into four cycles: intake, compression, power, and exhaust stages

The ratio of the volume of the gas in the cylinder after the first stroke to its volume after the second stroke is the *compression ratio* of the engine. Modern cars run at compression ratios of about 9:1, which means the gasoline-air mixture in the cylinder is compressed by a factor of nine in the second stroke. After the gasoline/air mixture is compressed, the spark plug at the top of the cylinder fires and the resulting explosion pushes the piston out of the cylinder in the third stroke. Finally, the piston is pushed back into the cylinder in the fourth stroke, clearing out the exhaust gases.

Liquids are much harder to compress than gases. They are so hard to compress that the hydraulic brake systems used in most cars operate on the principle that there is essentially no change in the volume of the brake fluid when pressure is applied to this liquid. Most solids are even harder to compress. The only exceptions belong to a rare class of compounds that includes natural and synthetic rubber. Most rubber balls that seem easy to compress, such as a racquetball, are filled with air, which is compressed when the ball is squeezed.

Expandability

Anyone who has walked into a kitchen where bread was baking has experienced the fact that gases expand to fill their containers, as the air in the kitchen becomes filled with wonderful odors. Unfortunately the same thing happens when someone breaks open a rotten egg and the characteristic odor of hydrogen sulfide (H_2S) rapidly diffuses through the room. Because gases expand to fill their containers, it is safe to assume that the volume of a gas is equal to the volume of its container.

Low Density

Gases contain scattered molecules that are dispersed across a given volume and are therefore less dense than in their solid or liquid states. Their low density gives gases fluidity, which allows gas particles to move rapidly and randomly past one another, expanding or contracting with no fixed positioning. The average distances between the molecules are large enough that interactions between molecules don't interfere with their motion.

Indefinite Shape or Volume

Gases have no definite shape or volume.The random movement of gas molecules allows them to expand or contract to assume the volume of the container holding them. Therefore, a gas's volume refers to the space of the container in which its molecules have range to move. This property results in gases occupying more space than they would in their liquid or solid state. Gases also contract and expand by predictable amounts depending upon changes in temperature and pressure.

Diffusivity

Given the large amounts of space between gas molecules, two or more gases can mix quickly and easily with one another to form a homogeneous mixture. This process is called diffusion.

The volume of a gas is one of its characteristic properties. Another characteristic property is the pressure the gas exerts on its surroundings. Many of us got our first exposure to the pressure of a gas when we rode to the neighborhood gas station to check the pressure of our bicycle tires. Depending on the kind of bicycle we had, we added air to the tires until the pressure gauge read between 30 and 70 pounds per square inch (lb/in^2 or psi). Two important properties of pressure can be obtained from this example.

1. The pressure of a gas becomes larger as more gas is added to the container.
2. Pressure is measured in units (such as lb/in^2) that describe the force exerted by the gas divided by the area over which this force is distributed.

The first conclusion can be summarized in the following relationship, where P is the pressure of the gas and n is the amount of gas in the container.

$$P \alpha n$$

Because the pressure increases as gas is added to the container, P is directly proportional to n.

The second conclusion describes the relationship between pressure and force. Pressure is defined as the force exerted on an object divided by the area over which the force is distributed.

$$Pressure = \frac{force}{area}$$

The difference between pressure and force can be illustrated with an analogy based on a 10-penny nail, a hammer, and a piece of wood shown in the figure below. By resting the nail on its point, and hitting the head with the hammer, we can drive the nail into the wood. But what happens if we turn the nail over and rest the head of the nail against the wood? If we hit the nail with the same force, we can't get the nail to stick into the wood.

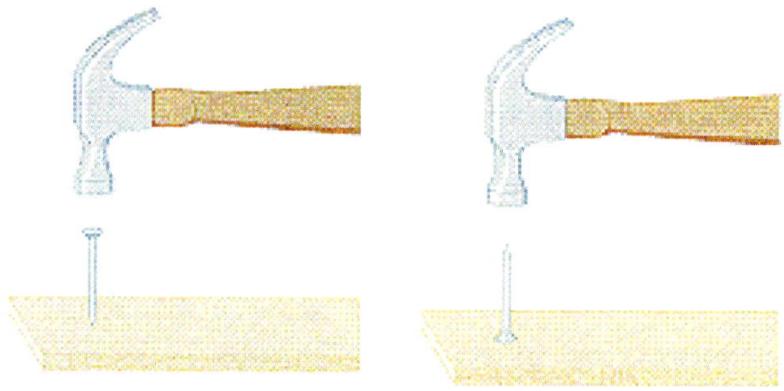

When we hit the nail on the head, the force of this blow is applied to the very small area of the wood in contact with the sharp point of the nail, and the nail slips easily into the wood. But when we turn the nail over, and hit it on the point, the force is distributed over a much larger area. The force is now distributed over the surface of the wood that touches any part of the nail head. As a result, the pressure applied to the wood is much smaller and the nail just bounces off the wood.

Ideal Gas

An ideal gas is one that conforms exactly to the tenets of the kinetic molecular theory, where the volume occupied by the gas particles is negligible relative to the total volume of the container, and there are no appreciable intermolecular attractions or repulsions.

The term ideal gas refers to a hypothetical gas composed of molecules which follow a few rules:

- Ideal gas molecules do not attract or repel each other. The only interaction between ideal gas molecules would be an elastic collision upon impact with each other or an elastic collision with the walls of the container.
- Ideal gas molecules themselves take up no volume. The gas takes up volume since the molecules expand into a large region of space, but the Ideal gas molecules are approximated as point particles that have no volume in and of themselves.

Real Gas

Real gases can deviate from ideal behaviour, especially at *high pressures* and *low temperatures*. The extent of deviation is measured using the compressibility factor. The compressibility factor is obtained by solving for n in the ideal gas law: dividing the product of pressure and volume by the product of the gas constant and temperature (PV/RT) for one mole of a given substance. Under ideal conditions, this ratio of PV/RT should be exactly equal to 1.

Real Gases at High Pressure

At higher pressures, gas molecules are closer together in a space. As a result of this crowding, gas molecules experience greater attractive intermolecular forces. Intermolecular forces hold molecules together more, lessening the force and frequency of collisions with the container wall and thus lowering the pressure below ideal values. As well, at higher pressure, molecules occupy a larger proportion of the volume of the container. With other gas molecules taking up a larger proportion of the volume of the container, the unoccupied volume of the container available to any one molecule is smaller than in ideal conditions. This decrease in available volume causes an increase in pressure beyond ideal conditions.

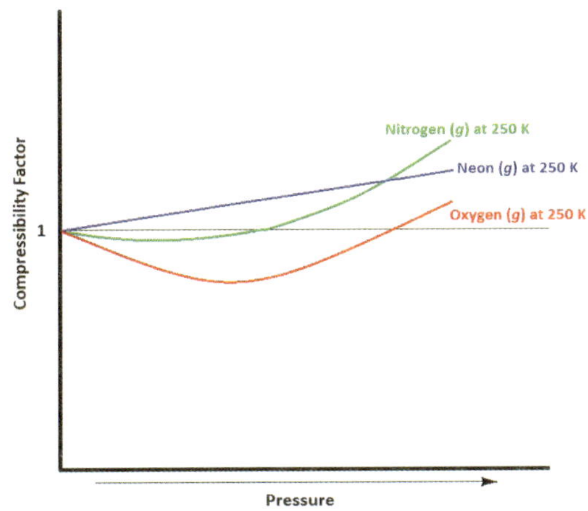

Approximate compressibility factors of three gases at 250 K

Real Gases at Low Temperature

Temperature also influences deviations from ideal gas behaviour. As temperature decreases, the average kinetic energy of the gas particles decreases. A larger proportion of gas molecules therefore have insufficient kinetic energy to overcome attractive intermolecular forces from neighbouring atoms. This means that gas molecules become "stickier" to each other, and collide with the walls of the container with less frequency and force, decreasing pressure below that of ideal values.

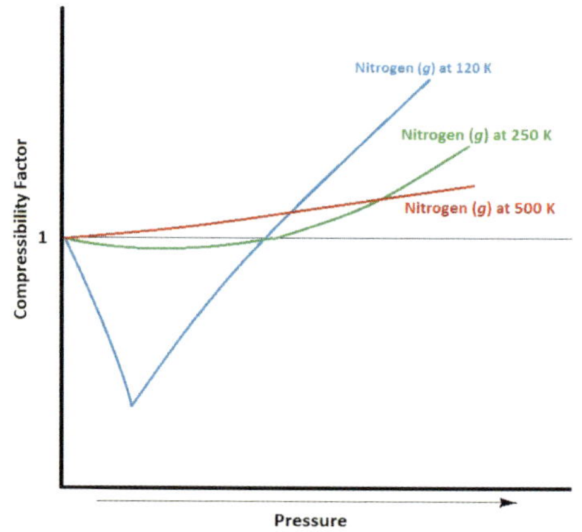

Approximate compressibility factor of nitrogen at different temperatures

Models

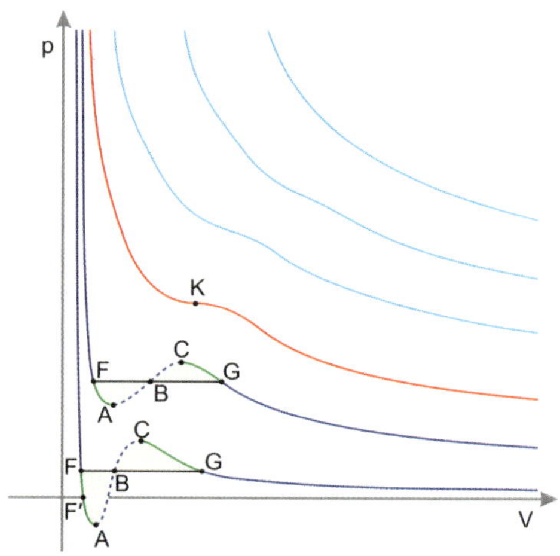

Isotherms of real gas

Dark blue curves – isotherms below the critical temperature. Green sections – metastable states.

The section to the left of point F – normal liquid.

Point F – boiling point.

Line FG – equilibrium of liquid and gaseous phases.

Section FA – superheated liquid.

Section F′A – stretched liquid ($p<0$).

Section AC – analytic continuation of isotherm, physically impossible.

Section CG – supercooled vapor.

Point G – dew point.

The plot to the right of point G – normal gas.

Areas FAB and GCB are equal.

Red curve – Critical isotherm.

Point K – critical point.

Light blue curves – supercritical isotherms

Van Der Waals Model

Real gases are often modeled by taking into account their molar weight and molar volume,

$$RT = \left(p + \frac{a}{V_{\mathrm{m}}^2}\right)(V_{\mathrm{m}} - b)$$

or alternatively:

$$p = \frac{RT}{V_m - b} - \frac{a}{V_m{}^2}$$

Where p is the pressure, T is the temperature, R the ideal gas constant, and V_{m} the molar volume. a and b are parameters that are determined empirically for each gas, but are sometimes estimated from their critical temperature (T_{c}) and critical pressure (p_{c}) using these relations:

$$a = \frac{27R^2T_{\mathrm{c}}^2}{64p_{\mathrm{c}}}$$

$$b = \frac{RT_{\mathrm{c}}}{8p_{\mathrm{c}}}$$

With the reduced properties $p_r = \frac{p}{p_{\mathrm{c}}}, V_r = \frac{V_{\mathrm{m}}}{V_{\mathrm{m,c}}}, T_r = \frac{T}{T_{\mathrm{c}}}$ the equation can be written in the *reduced form*:

$$p_r = \frac{8}{3}\frac{T_r}{V_r - \frac{1}{3}} - \frac{3}{V_r^2}$$

Redlich–Kwong Model

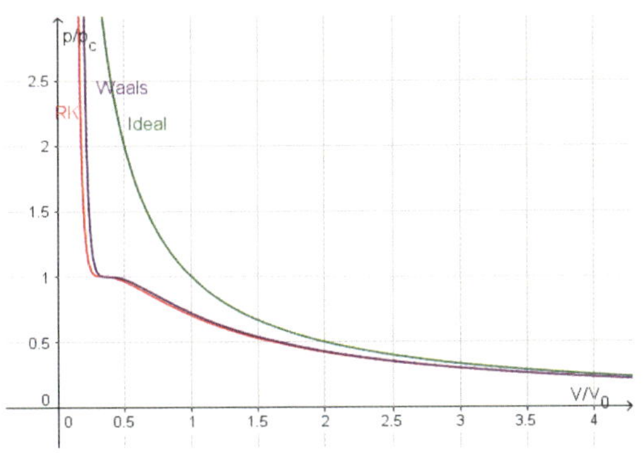

Critical isotherm for Redlich-Kwong model in comparison to van-der-Waals model and ideal gas (with $V_0 = RT_c / p_c$)

The Redlich–Kwong equation is another two-parameter equation that is used to model real gases. It is almost always more accurate than the van der Waals equation, and often more accurate than some equations with more than two parameters. The equation is,

$$RT = \left(p + \frac{a}{\sqrt{T} V_{\mathrm{m}} (V_{\mathrm{m}} + b)} \right) (V_{\mathrm{m}} - b)$$

or alternatively:

$$p = \frac{RT}{V_{\mathrm{m}} - b} - \frac{a}{\sqrt{T} V_{\mathrm{m}} (V_{\mathrm{m}} + b)}$$

where a and b two empirical parameters that are not the same parameters as in the van der Waals equation. These parameters can be determined:

$$= 0.42748 \frac{R^2 T_{\mathrm{c}}^{5/2}}{p_{\mathrm{c}}}$$

$$b = 0.08664 \frac{R T_{\mathrm{c}}}{p_{\mathrm{c}}}$$

Using $p_r = \frac{p}{p_{\mathrm{c}}}, V_r = \frac{V_{\mathrm{m}}}{V_{\mathrm{m,c}}}, T_r = \frac{T}{T_{\mathrm{c}}}$ the equation of state can be written in the *reduced form*:

$$p_r = \frac{3T_r}{V_r - b'} - \frac{1}{b' \sqrt{T_r} V_r (V_r + b')} \quad \text{with } b' = \sqrt[3]{2} - 1 \approx 0.26$$

Berthelot and Modified Berthelot Model

The Berthelot equation (named after D. Berthelot) is very rarely used,

$$p = \frac{RT}{V_{\mathrm{m}} - b} - \frac{a}{T V_{\mathrm{m}}^2}$$

but the modified version is somewhat more accurate,

$$p = \frac{RT}{V_{\mathrm{m}}} \left[1 + \frac{9p / p_{\mathrm{c}}}{128T / T_{\mathrm{c}}} \left(1 - \frac{6}{(T / T_{\mathrm{c}})^2} \right) \right]$$

Dieterici Model

This model (named after C. Dieterici) fell out of usage in recent years,

$$p = \frac{RT}{V_{\mathrm{m}} - b} \cdot \exp\left(-\frac{a}{V_{\mathrm{m}} RT} \right)$$

with parameters a,b and $\exp\left(-\frac{a}{V_m RT}\right)$ $\exp\left(-\frac{a}{V_m RT}\right) = e^{-\frac{a}{V_m RT}} = 1 - \frac{a}{V_m RT} + \ldots$

Clausius Model

The Clausius equation (named after Rudolf Clausius) is a very simple three-parameter equation used to model gases.

$$RT = \left(p + \frac{a}{T(V_m + c)^2}\right)(V_m - b)$$

or alternatively:

$$p = \frac{RT}{V_m - b} - \frac{a}{T(V_m + c)^2}$$

where,

$$a = \frac{27R^2T_c^3}{64p_c}$$

$$b = V_c - \frac{RT_c}{4p_c}$$

$$c = \frac{3RT_c}{8p_c} - V_c$$

where V_c is critical volume.

Virial Model

The Virial equation derives from a perturbative treatment of statistical mechanics.

$$pV = RT\left[+ \frac{B(T)}{V_m} + \frac{C(T)}{V_m^2} + \frac{D(T)}{V_m^3} + \ldots\right]$$

or alternatively:

$$pV_m = RT\left[1 + B'(T)p + C'(T)p^2 + D'(T)p^3 \ldots\right]$$

where A, B, C, A', B', and C' are temperature dependent constants.

Peng–Robinson Model

Peng–Robinson equation of state (named after D.-Y. Peng and D. B. Robinson) has the interesting property being useful in modeling some liquids as well as real gases.

$$p = \frac{RT}{V_m - b} - \frac{a(T)}{V_m(V_m + b) + b(V_m - b)}$$

Wohl Model

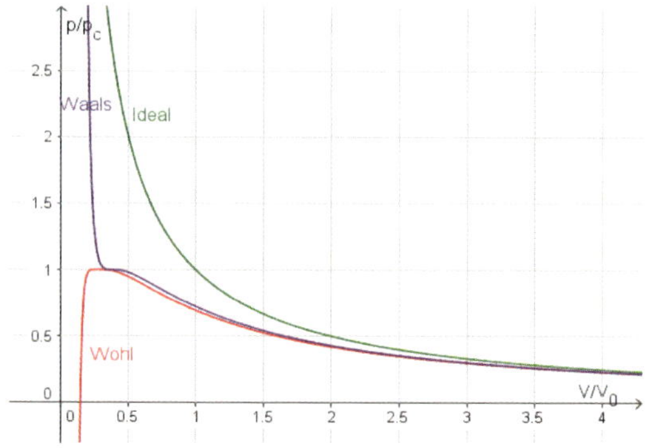

Isotherm ($V/V_0 -> p\_r$) at critical temperature for Wohl model, van der Waals model and ideal gas model (with $V_0 = RT_c / p_c$)

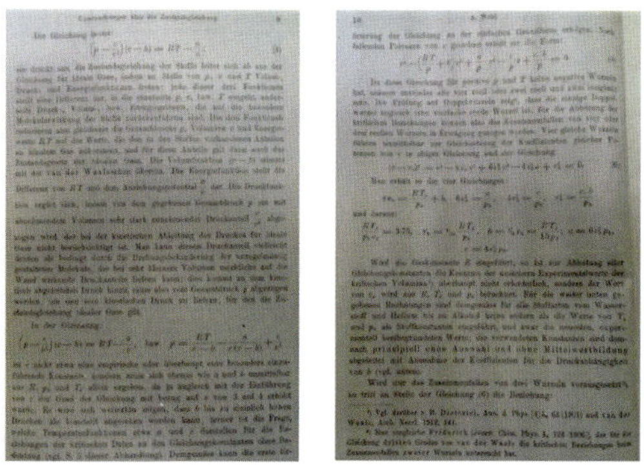

Untersuchungen über die Zustandsgleichung

The Wohl equation is formulated in terms of critical values, making it useful when real gas constants are not available, but it cannot be used for high densities, as for example the critical isotherm shows a drastic *decrease* of pressure when the volume is contracted beyond the critical volume.

$$p = \frac{RT}{V_m - b} - \frac{a}{TV_m(V_m - b)} + \frac{c}{T^2 V_m^3}$$

$$\text{or: } \left(p - \frac{c}{T^2 V_m^3}\right)(V_m - b) = RT - \frac{a}{TV_m}$$

or alternatively:

$$RT = \left(p + \frac{a}{TV_m(V_m - b)} - \frac{c}{T^2 V_m^3}\right)(V_m - b)$$

where,

$$a = 6p_c T_c V_{m,c}^2$$

$$b = \frac{V_{m,c}}{4} \text{ with } V_{m,c} = \frac{4}{15}\frac{RT_c}{p_c}$$

$c = 4p_c T_c^2 V_{m,c}^3$, where $V_{m,c}$, p_c, T_c are (respectively) the molar volume, the pressure and the temperature at the critical point.

And with the reduced properties $p_r = \frac{p}{p_c}, V_r = \frac{V_m}{V_{m,c}}, T_r = \frac{T}{T_c}$ one can write the first equation in the *reduced form*:

$$p_r = \frac{15}{4}\frac{T_r}{V_r - \frac{1}{4}} - \frac{6}{T_r V_r \left(V_r - \frac{1}{4}\right)} + \frac{4}{T_r^2 V_r^3}$$

Beattie–Bridgeman Model

This equation is based on five experimentally determined constants. It is expressed as,

$$p = \frac{RT}{\upsilon^2}\left(1 - \frac{c}{\upsilon T^3}\right)(\upsilon + B) - \frac{A}{\upsilon^2}$$

where,

$$A = A_0\left(1 - \frac{a}{\upsilon}\right)$$

$$B = B_0\left(1 - \frac{b}{\upsilon}\right)$$

This equation is known to be reasonably accurate for densities up to about 0.8 ρ_{cr}, where ρ_{cr} is the density of the substance at its critical point. The constants appearing in the above equation are available in the following table when p is in kPa, υ is in $\frac{m^3}{kmol}$, T is in K and R=8.314 $\frac{kPa \cdot m^3}{kmol \cdot K}$

Gas	A_0	a	B_0	b	c
Air	131.8441	0.01931	0.04611	-0.001101	4.34×10^4
Argon, Ar	130.7802	0.02328	0.03931	0.0	5.99×10^4
Carbon dioxide, CO_2	507.2836	0.07132	0.10476	0.07235	6.60×10^5
Helium, He	2.1886	0.05984	0.01400	0.0	40

Hydrogen, H_2	20.0117	-0.00506	0.02096	-0.04359	504
Nitrogen, N_2	136.2315	0.02617	0.05046	-0.00691	4.20×10^4
Oxygen, O_2	151.0857	0.02562	0.04624	0.004208	4.80×10^4

Benedict–Webb–Rubin Model

The BWR equation, sometimes referred to as the BWRS equation,

$$p = RTd + d^2\left(RT(B+bd) - (A + ad - a\alpha d^4) - \frac{1}{T^2}[C - cd(1+\gamma d^2)\exp(-\gamma d^2)]\right)$$

where d is the molar density and where a, b, c, A, B, C, α, and γ are empirical constants. Note that the γ constant is a derivative of constant α and therefore almost identical to 1.

Kinetic Theory of Gases

The Kinetic molecular theory provides the microscopic model of gases and gives us a picture of the behavior of gases. We have already had a look at the gas laws which depict the changing behavior of gases with changing physical conditions.

The question for the people studying these laws was to explain the cause of such behavior. As the findings of Kinetic theory of gases related to the behavior of molecules and atoms, this theory tries to answer such questions. It relates the physical changes in a gas with the microscopic behavior of the gases.

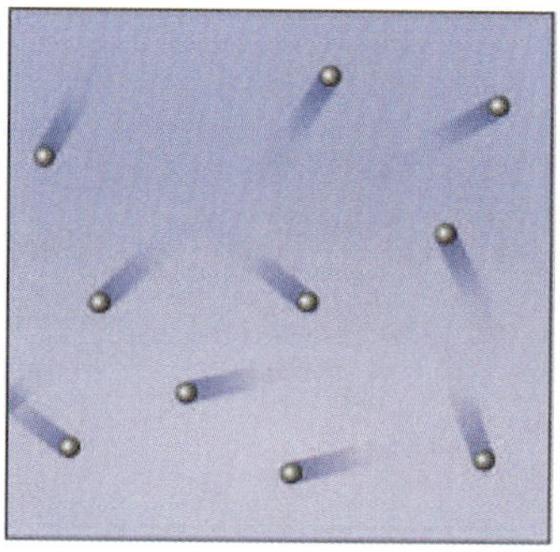

Postulates of the Theory

Kinetic Molecular theory intensely outlines the molecular behavior of a gas. The theory is based on the following postulates.

Negligible Molecules to Space-volume Ratio

Gases consist of a large number of atoms or molecules that are generally identical in shape and size and are spaced very far apart. Due to the large space between these molecules, the volume of molecules is negligible as compared to the space that a gas occupies. The space between these molecules is empty which explains the high compressibility of gases.

No Force of Attraction

At normal temperatures and pressures, gaseous molecules generally lack attraction forces. You might have noticed that a gas occupies the whole space, irrespective of the shape of the container. It is when the temperature decreases and the pressure increases, that the force of attraction between the molecules starts to build.

Now how do we say that the molecules in gases have almost nil force of attraction? The property of gases to occupy the available space instantly proves this assumption.

Always in Motion

A large amount of space between the molecules gives them the room to be always in motion. Gases do not have a fixed shape, this shows that the gaseous molecules are always in a random motion. The scattered shape filling behavior of the gaseous element signifies the constantly moving gas molecules.

The molecules of gases move in a straight line, in all directions. And with such random motion, these molecules are bound to collide with one another. The highly mobile molecules in their course of motion collide with the walls of the container, thus exerting pressure on these walls.

Elastic Collision

The collision between the molecules in a gas is perfectly elastic. Collisions are of two types: Elastic and Inelastic. In an elastic collision, the energy stored in the molecules by virtue of their movement is same before and after the collision. If the collisions were inelastic, after some time duration, the molecules would come to a rest. We do not observe this so the collisions must be elastic in nature.

Constantly Changing Kinetic Energy

The Kinetic energy stored in the various molecules is never static or constant. It is always changing. The kinetic energy of an object is by virtue of its motion or movement. Since the different molecules or atoms in gases are always in a random motion they move with different speed. Each molecule with a changing speed of its own shows changing levels of kinetic energy.

The changing speed of molecules is also because these particles randomly move in any direction, thence collide. As soon as the collision occurs it is obvious that the speed changes. This results in the changing Kinetic energy of the gaseous element. Now if we presume that the initial speed of the particles is same, yet due to the ever- repeating collision between these molecules the final speed changes constantly. The changed speed in these gaseous molecules results in the changed Kinetic energy level.

Average Kinetic Energy

Like the speed of a moving car with no definite speed, molecules and gas atoms of a gas show a similar behavior in terms of Kinetic energy. The never-ending, movement of molecules and particles results in a changed Kinetic Energy. Since for calculating the kinetic energy, we use the equation: K.E $=1/2\ mv^2$, and with the changing speed, the K.E is bound to change.

Kinetic energy like the speed of the car is measured in the form of Average Kinetic energy. The variable speed gives a variable K.E, so for matters of high precision, we use average kinetic energy to keep a track over the changing scenario of the energy system.

Effects of Physical Conditions

According to the Kinetic theory of gases, the kinetic energy of the molecules is directly proportional to temperature. The temperature here is taken as the absolute temperature. At constant volume, the pressure of the gas increases with increasing temperature. This is because with increasing temperature, the kinetic energy of the molecules increases.

The increased kinetic energy of the molecules makes them strike with the walls of the container more often, hence leading to an increased pressure on the walls of the container.

From the postulates of Kinetic Molecular Theory, we now understand the basic nature of molecules in the gaseous state. Kinetic theory supports the various gas laws. The Kinetic theory of gases helps in interpreting the experimental observations of the gas laws.

Gas Laws

Laws which describe properties of gases with respect to temperature, volume, pressure, and moles are known as gas laws. There are different types of gas laws, which were proposed in different periods by different scientists.

1. Avogadro's Law
2. Boyle's law
3. Charles's Law
4. Gay–Lussac law
5. Ideal gas law

Avogadro's Law

Avogadro's Law (Avogadro's theory; Avogadro's hypothesis) is a principle stated in 1811 by the Italian chemist Amedeo Avogadro (1776-1856) that "equal volumes of gases at the same temperature and pressure contain the same number of molecules regardless of their chemical nature and physical properties". This number (Avogadro's number) is 6.022×10^{23}. It is the number of molecules of any gas present in a volume of 22.41 L and is the same for the lightest gas (hydrogen) as for a heavy gas such as carbon dioxide or bromine.

The law can be stated mathematically,

$$\frac{V}{n} = k.$$

where:

V is the volume of the gas.

n is the amount of substance of the gas.

k is a proportionality constant.

The most important consequence of Avogadro's law is that the ideal gas constant has the same value for all gases. This means that the constant,

$$\frac{p_1 \cdot V_1}{T_1 \cdot n_1} = \frac{p_2 \cdot V_2}{T_2 \cdot n_2} \text{ constant}$$

where:

p is the pressure of the gas.

T is the temperature of the gas.

has the same value for all gases, independent of the size or mass of the gas molecules.

One mole of an ideal gas occupies 22.4 liters (dm^3) at STP, and occupies 24.45 litres at SATP (Standard Ambient Temperature and Pressure = 273K and 1 atm or 101.325 kPa). This volume is often referred to as the molar volume of an ideal gas. Real gases may deviate from this value.

Or to put it another way "the principle that equal volumes of all gases at the same temperature and pressure contain the same number of molecules. Thus, the molar volume of all ideal gases at 0°C and a pressure of 1 atm. is 22.4 liters"

Avogadro's number is one of the fundamental constants of chemistry. It permits calculation of the amount of pure substance (mole), the basis of stoichiometric relationships. It also makes possible determination of how much heavier a simple molecule of one gas is than that of another, as a result the relative molecular weights of gases can be ascertained by comparing the weights of equal volumes.

Avogadro's number (conventionally represented by N' in chemical calculations) is now considered to be the number of atoms present in *12 grams* of the carbon-12 isotope (one mole of carbon 12) and can be applied to any type of chemical entity.

Implications of Avogadro's Law

There are a few important consequences of the law being true.

- The molar volume of all ideal gases at 0°C and 1 atm pressure is 22.4 liters.
- If pressure and temperature of a gas are constant, when the amount of gas increases, the volume increases.
- If pressure and temperature of a gas are constant, when the amount of gas decreases, the volume decreases.
- You prove Avogadro's Law every time you blow up a balloon.

Avogadro's Law Example

Say you have 5.00 L of a gas which contains 0.965 mol of molecules. What will be the new volume of the gas if the quantity is increased to 1.80 mol, assuming pressure and temperature are held constant?

Select the appropriate form of the law for the calculation.

In this case, a good choice is:

$$V_1 n_2 = V_2 n_1$$

$$(5.00\ \text{L})(1.80\ \text{mol}) = (x)(0.965\ \text{mol})$$

Rewriting to solve for x give you:

x = (5.00 L)(1.80 mol) / (0.965 mol)

x = 9.33 L

Boyle's Law

On Earth, matter exists in one of three states: solid, liquid or gas. Matter in each state exhibits distinct characteristics. Gases, for example, do not have a fixed volume* or shape. As a result, gases respond to pressure changes by changing their volume. In other words, gases are compressible. In contrast, liquids and solids are not compressible: their volume does not change in response to changing pressure. This is the reason air-filled spaces in our ears "pop" during airplane takeoff and landings while the liquid filled spaces in our bodies do not. Boyle's Law* describes the relationship between pressure and volume at constant temperature for a fixed mass* (number of molecules) of a gas.

To understand Boyle's law, it helps to visualize the behavior of gas particles (or molecules) in an enclosed space. In a closed container, individual molecules are constantly hitting and bouncing off the container walls. Each time a gas molecule bounces, it imparts a force on the wall. In a flexible container such as a balloon, the force of the molecules hitting the inside of the balloon hold the balloon inflated. The force of each impact is small, but the sheer number of collisions create enough force to keep the balloon open.

Pressure in a closed container changes if,

1. Temperature changes
2. Number of molecules increases or decreases
3. Volume changes.

Boyle's Law deals with number 3; the relationship between volume and pressure when the other two remain constant.

According to Boyle's Law, the amount a gas will compress is proportional to the pressure applied. Its mathematical expression is:

$$P_1V_1 = P_2V_2$$

Where, P_1 is the pressure of a quantity of gas with a volume of V_1 and P_2 is the pressure of the same quantity of gas when it has a volume of V_2. This means that if nothing else changes, the volume of a given mass of gas is inversely proportional to pressure it is under. It is a linear relationship. If pressure on a gas doubles, its volume will decrease by 1/2. An alternative expression of the law is:

$$PV = C$$

The product of the volume and pressure equals a constant.

The relationship between pressure and volume results from the influence volume has on the rate at which gas molecules collide with the container walls. If the volume decreases, the molecules encounter the container walls more often. This is true even though the speed (temperature) of the individual molecules has not changed. An increased number of collisions equates to higher pressure. Conversely, if volume increases, the rate of collisions and the pressure decreases.

Charles's Law

$$V \propto T$$

This relationship between the temperature and volume of a gas, which became known as Charles' law, provides an explanation of how hot-air balloons work. Ever since the third century B.C., it has been known that an object floats when it weighs less than the fluid it displaces. If a gas expands when heated, then a given weight of hot air occupies a larger volume than the same weight of cold air. Hot air is therefore less dense than cold air. Once the air in a balloon gets hot enough, the net weight of the balloon plus this hot air is less than the weight of an equivalent volume of cold air, and the balloon starts to rise. When the gas in the balloon is allowed to cool, the balloon returns to the ground.

Charles' law can be demonstrated with the apparatus shown in the figure below. A 30-mL syringe and a thermometer are inserted through a rubber stopper into a flask that has been cooled to 0° C. The ice bath is then removed and the flask is immersed in a warm-water bath. The gas in the flask expands as it warms, slowly pushing the piston out of the syringe. The total volume of the gas in the system is equal to the volume of the flask plus the volume of the syringe.

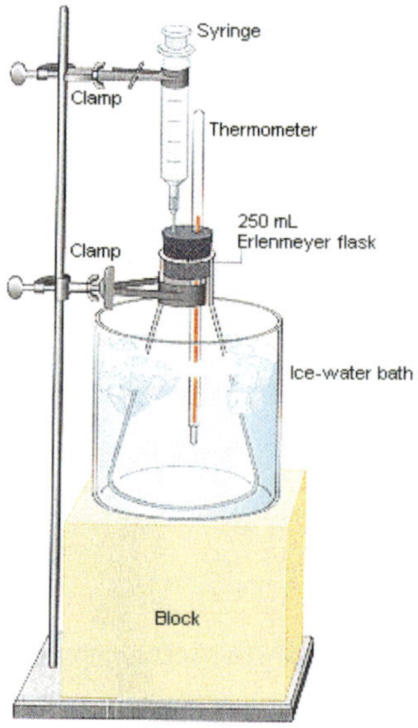

The figure shows a plot of the typical data obtained from this experiment

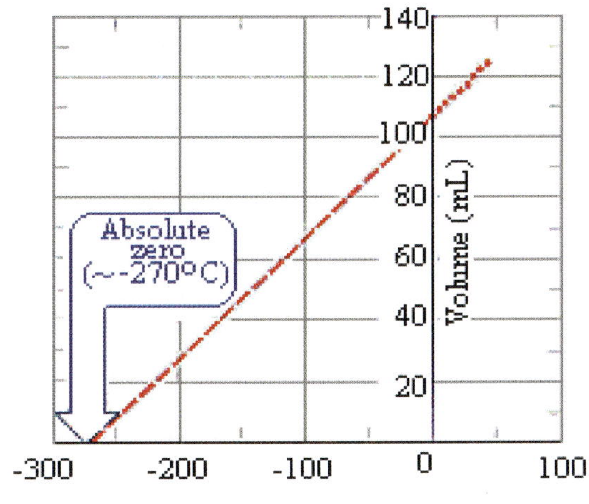

This graph provides us with another way of defining absolute zero on the temperature scale. Absolute zero is the temperature at which the volume of a gas becomes zero when the a plot of the volume versus temperature for a gas are extrapolated. As expected, the value of absolute zero obtained by extrapolating the data is essentially the same as the value obtained from the graph of pressure versus temperature in the preceding section. Absolute zero can therefore be more accurately defined as the temperature at which the pressure and the volume of a gas extrapolate to zero.

A plot of the volume versus the temperature of a gas (when the temperatures obtained are converted from Celsius to the Kelvin scale) becomes a straight line that passes through the origin. Any two points along this line can therefore be used to construct the following equation, which is known as Charles' law.

$$\frac{V_1}{V_2} = \frac{T_1}{T_2}$$

Relation to Absolute Zero

Charles' law appears to imply that the volume of a gas will descend to zero at a certain temperature (−266.66°C according to Gay-Lussac's figures) or −273.15°C. Gay-Lussac was clear in his description that the law was not applicable at low temperatures:

but I may mention that this last conclusion cannot be true except so long as the compressed vapours remain entirely in the elastic state; and this requires that their temperature shall be sufficiently elevated to enable them to resist the pressure which tends to make them assume the liquid state.

At absolute zero temperature the gas possesses zero energy and hence the molecules restrict motion. Gay-Lussac had no experience of liquid air (first prepared in 1877), although he appears to have believed (as did Dalton) that the "permanent gases" such as air and hydrogen could be liquified. Gay-Lussac had also worked with the vapours of volatile liquids in demonstrating Charles' law, and was aware that the law does not apply just above the boiling point of the liquid:

> I may however remark that when the temperature of the ether is only a little above its boiling point, its condensation is a little more rapid than that of atmospheric air. This fact is related to a phenomenon which is exhibited by a great many bodies when passing from the liquid to the solid state, but which is no longer sensible at temperatures a few degrees above that at which the transition occurs.

The first mention of a temperature at which the volume of a gas might descend to zero was by William Thomson (later known as Lord Kelvin) in 1848:

> This is what we might anticipate, when we reflect that infinite cold must correspond to a finite number of degrees of the air-thermometer below zero; since if we push the strict principle of graduation, stated above, sufficiently far, we should arrive at a point corresponding to the volume of air being reduced to nothing, which would be marked as −273° of the scale (−100/.366, if .366 be the coefficient of expansion); and therefore −273° of the air-thermometer is a point which cannot be reached at any finite temperature, however low.

However, the "absolute zero" on the Kelvin temperature scale was originally defined in terms of the second law of thermodynamics, which Thomson himself described in 1852. Thomson did not assume that this was equal to the "zero-volume point" of Charles' law, merely that Charles' law provided the minimum temperature which could be attained.

However, Charles also stated:

> The volume of a fixed mass of dry gas increases or decreases by $1/_{273}$ times the volume at 0°C for every 1°C rise or fall in temperature. Thus:
>
> $$V_T = V_0 + (\tfrac{1}{273} \times V_0) \times T$$
>
> $$V_T = V_0 (1 + \tfrac{T}{273})$$
>
> where V_T is the volume of gas at temperature T, V_0 is the volume at 0°C.

Relation to Kinetic Theory

The kinetic theory of gases relates the macroscopic properties of gases, such as pressure and volume, to the microscopic properties of the molecules which make up the gas, particularly the mass and speed of the molecules. In order to derive Charles' law from kinetic theory, it is necessary to have a microscopic definition of temperature: this can be conveniently taken as the temperature being proportional to the average kinetic energy of the gas molecules, $\overline{E}_k$:

$$T \propto \overline{E}_k .$$

Under this definition, the demonstration of Charles' law is almost trivial. The kinetic theory equivalent of the ideal gas law relates *PV* to the average kinetic energy:

$$PV = \frac{2}{3}N\bar{E}_k$$

Gay-Lussac's Law

Gay-Lussac's most important contributions to the study of gases, however, were experiments he performed on the ratio of the volumes of gases involved in a chemical reaction.

Gay-Lussac studied the volume of gases consumed or produced in a chemical reaction because he was interested in the reaction between hydrogen and oxygen to form water. He argued that measurements of the *weights* of hydrogen and oxygen consumed in this reaction could be influenced by the moisture present in the reaction flask, but this moisture would not affect the *volumes* of hydrogen and oxygen gases consumed in the reaction.

Much to his surprise, Gay-Lussac found that 199.89 parts by volume of hydrogen were consumed for every 100 parts by volume of oxygen. Thus, hydrogen and oxygen seemed to combine in a simple 2:1 ratio by volume.

hydrogen	+	oxygen	→	water
2 volumes		1 volume		

Gay-Lussac found similar whole-number ratios for the reaction between other pairs of gases. The compound we now know as hydrogen chloride (HCl) combined with ammonia (NH_3) in a simple 1:1 ratio by volume:

hydrogen chloride	+	ammonia	→	ammonium chloride
1 volume		1 volume		

Carbon monoxide combined with oxygen in a 2:1 ratio by volume:

carbon monoxide	+	oxygen	→	carbon dioxide
2 volumes		1 volume		

Gay-Lussac obtained similar results when he analyzed the volumes of gases given off when compounds decomposed. Ammonia, for example, decomposed to give three times as much hydrogen by volume as nitrogen:

ammonia	→	nitrogen	+	hydrogen
		1 volume		3 volumes

On 31 December 1808, Gay-Lussac announced his law of combining volumes to a meeting of the Societ Philomatique in Paris. At that time, he summarized the law as follows: Gases combine among themselves in very simple proportions. Today, Gay-Lussac's law is stated as follows: The ratio of the volumes of gases consumed or produced in a chemical reaction is equal to the ratio of simple whole numbers.

Gay Lussacs Law Formula

Gay Lussac's law defines the relationship between the pressure and temperature of a gas. According to the law, when volume is held constant, the pressure of an ideal gas is directly proportional to its temperature. This law can be expressed as follows:

$$\frac{P_1}{T_1} = \frac{P_2}{T_2}$$

Where, P1 and P2 are the Initial Pressure and Final Pressure respectively. Both the pressures are in the units of atm. T1 refers to the Initial Temperature and T2 refers to the Final Temperature. The temperatures are measured in the units of Kelvin.

This can be explained as, at constant volume of a close container, if the pressure inside container increases, the temperature inside the container also increases. Pressure vs. Temperature graph is linear for ideal gases.

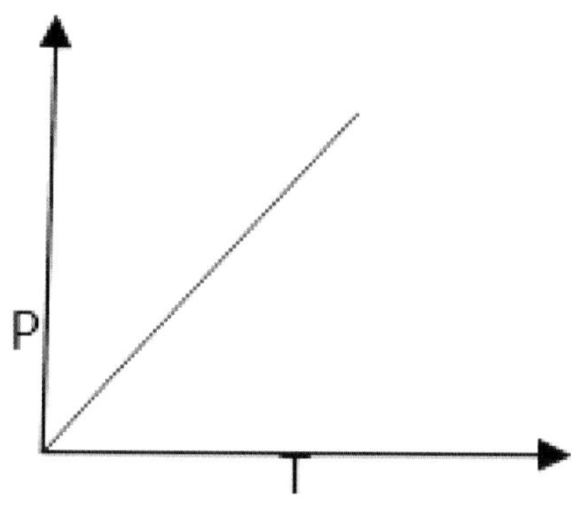

Gay Lussac's Law in Real Life

Gay Lussac's law has a wide range of applications in our daily life. Some of the common examples are given below:

- A burning automobile tire: Burning rubber will cause the increase in the air pressure in the tire due to increased temperature. This may lead to explosion of tire wall.
- Firing a bullet: Gay Lussac's law can be applied in firing a bullet. Here, the ignition of gunpowder generates a superheated gas which in turn increases the pressure so that the bullet travels longer with high speed.

- Heating a closed aerosol can: When you burn aerosol cans like spray paint and deodorants, the increased temperature can increase the pressure which may cause the explosion of the container. This is why there is warning on the deodorant bottles ‘not to put even empty bottles in fire’.
- Heating food in an oven: When you keep food in an oven for heating, the air inside the oven is heated, thus pressurized.

Solved Examples:

Question 1: At temperature 250K, a gas in a cylinder has a pressure of 1.5 atm. If the initial pressure of the gas was 1 atm, what was its initial temperature?

Solution: We have the variables,

Initial Pressure P1 = 1atm

Final Pressure P2 = 1.5 atm

Final Temperature T2 =250K

Initial Temperature T1 =?

Applying Gay Lussac’s law we get,

$$T_1 = \frac{P_1 T_2}{P_2}$$

T1= (1 x 250) / 1.5 = 166.6 K

Question 2: A gas container is at a temperature of 150K and a pressure of 1520 torr. The gas is heated to 300K. Find the new pressure in the unit of atm.

Solution: Given,

Initial Pressure P1= 1520 torr

Final Pressure P2 = ?

Initial Temperature T1 = 150

Final Temperature T2 = 300

For converting pressure in torr to atm, we use the relation:

1 atm = 760 torr

∴ 1520720 = 2 atm

Using Gay Lussac’s Law Formula,

$$P_1 T_2 = P_2 T_1$$

$$P_2 = \frac{P_1 T_2}{T_1}$$

$$P_2 = \frac{2 \times 300}{150} = 4\ atm$$

Ideal Gas Law

Because gases have common behaviors described by the gas laws, we can understand and predict the behavior of real gases through the concept of an ideal gas—a theoretical, idealized gas that always behaves according to the ideal gas equation.

The ideal gas equation is derived from the gas laws. This equation describes the relationships between all of the variables examined in the gas laws: pressure (*P*), volume (*V*), amount (*n*, in moles), and absolute temperature (*T*, in Kelvins). Along with the gas constant, *R*, these variables combine into the ideal gas equation:

$$PV = nRT$$

Using the ideal gas equation, we can solve for any one of the unknown variables, so long as we know the others. The value for *R* depends on the units used for the other variables.

Values of the gas constant *R* for different units.

Units	R's Value
cal K^{-1} mol^{-1}	1.9872
J K^{-1} mol^{-1}	8.3145
L atm K^{-1} mol^{-1}	0.0821
L Torr K^{-1} mol^{-1}	62.364
Pa m^3 K^{-1} mol^{-1}	8.3145

The ideal gas law assumes that an ideal gas's molecules have no volume, and experience no intermolecular attraction or repulsion. But real gas molecules do have finite volume and often have some (very small) interactions with each other. Nonetheless, the behavior and state of a real gas can often be predicted from the ideal gas equation, especially at standard temperature and pressure. Under most conditions the difference between a real gas's behavior and an ideal gas's behavior is so small that we can use the ideal gas equation for real gases.

Molar form of the Ideal Gas Law

The pressure, P, volume V, and temperature T of an ideal gas are related by a simple formula called the ideal gas law. The simplicity of this relationship is a big reason why we typically treat gases as ideal, unless there is a good reason to do otherwise.

$$PV = nRT$$

Where P is the pressure of the gas, V is the volume taken up by the gas, T is the temperature of the gas, RRR is the gas constant, and n is the number of moles of the gas.

Perhaps the most confusing thing about using the ideal gas law is making sure we use the right units when plugging in numbers. If you use the gas constant $R = 8.31K \frac{J}{K \cdot mol}$ then you must plug in the pressure P in units of pascals P_a, volume V in units of m^3, and temperature T in units of kelvin K.

If you use the gas constant $R = 0.082 \frac{L \cdot atm}{K \cdot mol}$ then you must plug in the pressure P in units of atmospheres atm, volume V in units of liters L and temperature T in units of kelvin K.

This information is summarized for convenience in the chart below.

Units to use for $PV = nRT$

$R = 8.31K \frac{J}{K \cdot mol}$	$R = 0.082 \frac{L \cdot atm}{K \cdot mol}$
Pressure in pascals *Pa*	Pressure in atmospheres *atm*
Volume in m^3	volume in liters *L*
Temperature in kelvin *K*	Temperature in kelvin *K*

Molecular form of the Ideal Gas Law

If we want to use N number of molecules instead of n moles, we can write the ideal gas law as,

$$PV = Nk_B T$$

Where P is the pressure of the gas, V is the volume taken up by the gas, T is the temperature of the gas, N is the number of molecules in the gas, and k_B is Boltzmann's constant,

$$k_B = 1.38 \times 10^{-23} \frac{J}{K}$$

When using this form of the ideal gas law with Boltzmann's constant, we have to plug in pressure P in units of pascals Pa, volume V in m^3, and temperature T in kelvin K. This information is summarized for convenience in the chart below.

Units to use for $PV = Nk_BT$ $k_B = 1.38 \times 10^{-23}$
Pressure in pascals Pa
Volume in m^3
Temperature in kelvin K

Proportional form of the Ideal Gas Law

There's another really useful way to write the ideal gas law. If the number of moles n (i.e. molecules N) of the gas doesn't change, then the quantity nR and Nk_B are constant for a gas. This happens frequently since the gas under consideration is often in a sealed container. So, if we move the pressure, volume and temperature onto the same side of the ideal gas law we get,

$$nR = Nk_B = \frac{PV}{T} = constant$$

This shows that, as long as the number of moles (i.e. molecules) of a gas remains the same, the quantity $\frac{PV}{T}$ is constant for a gas regardless of the process through which the gas is taken. In other words, if a gas starts in state 1 (with some value of pressure P_1, volume V_1, and temperature T_1) and is altered to a state 2 (with P_2 , volume V_2, and temperature T_2), then regardless of the details of the process we know the following relationship holds.

$$\frac{P_1V_1}{T_1} = \frac{P_2V_2}{T_2}$$

This formula is particularly useful when describing an ideal gas that changes from one state to another. Since this formula does not use any gas constants, we can use whichever units we want, but we must be consistent between the two sides (e.g. if we use m^3 for V_1 we'll have to use m^3 for V_2).

Combined Gas Law

The combined gas law combines the three gas laws: Boyle's Law, Charles' Law, and Gay-Lussac's Law. It states the ratio of the product of pressure and volume and the absolute temperature of a gas is equal to a constant. When Avogadro's law is added to the combined gas law, the ideal gas law results. Unlike the named gas laws, the combined gas law doesn't have an official discoverer.

It is simply a combination of the other gas laws that works when everything except temperature, pressure, and volume are held constant.

There are a couple of common equations for writing the combined gas law. The classic law relates Boyle's law and Charles' law to state:

$$PV/T = k$$

where

P = pressure

V = volume

T = absolute temperature (Kelvin)

k = constant

The constant k is a true constant if the number of moles of the gas doesn't change, otherwise it varies.

Another common formula for the combined gas law relates "before and after" conditions of a gas:

$$P_1V_1 / T_1 = P_2V_2 / T_2$$

Combined Gas Law Example

Find the volume of a gas at STP when 2.00 liters is collected at 745.0 mm Hg and 25.0°C.

To solve the problem, you first need to identify which formula to use. In this case, the question asks about conditions at STP, so you know you're dealing with a "before and after" problem. Next, you need to now what STP is.

If you haven't memorized this already (and you probably should, since it appears a lot), STP refers to "standard temperature and pressure", which is 273 K and 760.0 mm Hg.

Because the law works using absolute temperature, you need to convert 25.0°C to the Kelvin scale. This gives you 298 K.

At this point, you can just plug the values into the formula and solve for the unknown, but a common mistake when you're new to this type of problem is confusing which numbers go together.

It's good practice to identify the variables. In this problem:

$P_1 = 745.0$ mm Hg

$V_1 = 2.00$ L

$T_1 = 298$ K

$P_2 = 760.0$ mm Hg

$V_2 = x$ (the unknown you're solving for)

$T_2 = 273$ K

Next, take the formula and set it up to solve for your "x", which is V_2 in this problem.

$$P_1V_1 / T_1 = P_2V_2 / T_2$$

Cross-multiply to clear the fractions:

$$P_1V_1T_2 = P_2V_2T_1$$

Divide to isolate V_2:

$$V_2 = (P_1V_1T_2) / (P_2T_1)$$

Plug in the numbers:

$$V_2 = (745.0 \text{ mm Hg} \cdot 2.00 \text{ L} \cdot 273 \text{ K}) / (760 \text{ mm Hg} \cdot 298 \text{ K})$$

$$V_2 = 1.796 \text{ L}$$

Report the value using the correct number of significant figures:

$$V_2 = 1.80 \text{ L}$$

Uses of the Combined Gas Law

The combined gas law has practical applications when dealing with gases at ordinary temperatures and pressures. Like other gas laws based in ideal behavior, it becomes less accurate at high temperatures and pressures. The law is used in thermodynamics and fluid mechanics. For example, it can be used to calculate pressure, volume, or temperature for the gas in refrigerators or in clouds to forecast weather.

References

- Hsieh, Jui Sheng, Engineering Thermodynamics, Prentice-Hall Inc., Englewood Cliffs, New Jersey 07632, 1993. ISBN 0-13-275702-8

- What-is-the-ideal-gas-law, temp-kinetic-theory-ideal-gas-law, physics, science: khanacademy.org, Retrieved 28 March 2018
- Peng, D. Y. & Robinson, D. B. (1976). "A New Two-Constant Equation of State". Industrial and Engineering Chemistry: Fundamentals. 15: 59–64. doi:10.1021/i160057a011
- Gas-state-of-matter, science: britannica.com, Retrieved 19 June 2018
- Stanley M. Walas, Phase Equilibria in Chemical Engineering, Butterworth Publishers, 1985. ISBN 0-409-95162-5
- Properties-of-Gases, library, Chemistry: visionlearning.com, Retrieved 2 July 2018
- Yunus A. Cengel and Michael A. Boles, Thermodynamics: An Engineering Approach 7th Edition, McGraw-Hill, 2010, ISBN 007-352932-X

5

Electrochemistry

Electrochemistry is a sub-discipline of physical chemistry that studies the inter-relationship between electricity and chemical change, with one being the cause or effect of the other. The topics elaborated in this chapter, such as oxidation and reduction, corrosion and electrolysis, will be vital for a comprehensive understanding of the subject. It also discusses in detail, the different types of electrochemical cells such as galvanic cell and electrolytic cell.

Electrochemistry deals with the links between chemical reactions and electricity. This includes the study of chemical changes caused by the passage of an electric current across a medium, as well as the production of electric energy by chemical reactions. Electrochemistry also embraces the study of electrolyte solutions and the chemical equilibria that occur in them.

Many chemical reactions require the input of energy. Such reactions can be carried out at the surfaces of electrodes in cells connected to external power supplies. These reactions provide information about the nature and properties of the chemical species contained in the cells, and can also be used to synthesize new chemicals. The production of chlorine and aluminum and the electroplating and electrowinning of metals are examples of industrial electrochemical processes. Electrochemical cells that produce electric energy from chemical energy are the basis of primary and secondary (storage) batteries and fuel cells. Other electrical phenomena of interest in chemical systems include the behavior of ionic solutions and the conduction of current through these solutions, the separation of ions by an electric field (electrophoresis), the corrosion and passivation of metals, electrical effects in biological systems (bioelectrochemistry), and the effect of light on electrochemical cells (photoelectrochemistry).

Applications of Electrochemistry

Electrochemistry has a number of different uses, particularly in industry. The principles of cells are used to make electrical batteries. In science and technology, a battery is a device that stores chemical energy and makes it available in an electrical form. Batteries are made of electrochemical devices such as one or more galvanic cells or fuel cells. Batteries have many uses including in:

- Torches.
- Electrical appliances such as cellphones (long-life alkaline batteries).
- Digital cameras (lithium batteries).
- Hearing aids (silver-oxide batteries).
- Digital watches (mercury/silver-oxide batteries).
- Military applications (thermal batteries).

Electroplating

The electrolytic cell can be used for electroplating.

Electroplating occurs when an electrically conductive object is coated with a layer of metal using electrical current. Sometimes, electroplating is used to give a metal particular properties or for aesthetic reasons:

- Corrosion protection.
- Abrasion and wear resistance.
- The production of jewellery.

a) An electroplated piece of aluminium artwork and b) a wax stool electroplated in copper

Electro-refining (also sometimes called electrowinning) is electroplating on a large scale. Copper plays a major role in the electrical industry as it is very conductive and is used in electric cables. One of the problems though is that copper must be pure if it is to be an effective current carrier. One of the methods used to purify copper, is electrowinning (copper ore is processed into impure *blister copper*, which is then deposited as pure copper through electroplating). The copper electrowinning process is as follows:

1. A bar of impure copper containing other metallic impurities acts as the anode.
2. The cathode is made up of pure copper with few impurities.
3. The electrolyte is a solution of aqueous $CuSO_4$ $CuSO_4$ and H_2 SO_4 H_2SO_4.

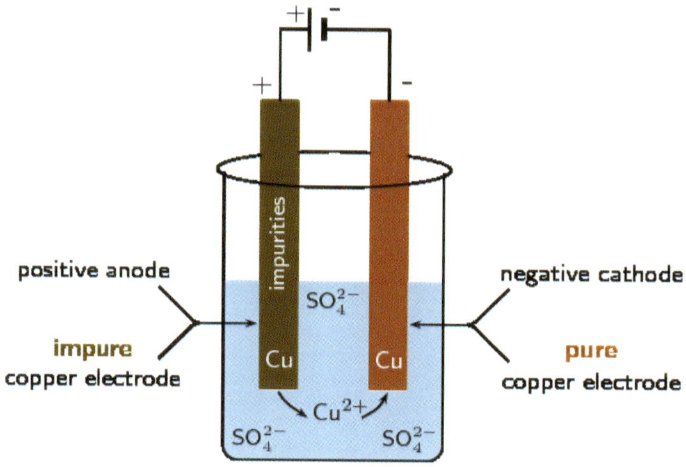

A simplified diagram to illustrate what happens during the electrowinning of copper

1. When current passes through the cell, electrolysis takes place:
    - The impure copper anode oxidises to form Cu^{2+} ions in solution. The anode decreases in mass in mass.

      $Cu(s) \rightarrow Cu^{2+}(aq) + 2e^-$

    - At the cathode reduction of positive copper ions takes place to produce pure copper metal. The cathode increases in mass.

      $Cu^{2+}(aq) + 2e^- \rightarrow Cu(s)$ (> 99% purity)

2. The other metal impurities do not dissolve (Au(s), Ag(s)) and form a solid sludge at the bottom of the tank or remain in solution (Zn(aq), Fe(aq) and Pb(aq)) in the electrolyte.

The Chloralkali Industry

The chlorine-alkali (chloralkali) industry is an important part of the chemical industry, which produces chlorine and sodium hydroxide through the electrolysis of the raw material brine. Brine is a saturated solution of sodium chloride (NaCl) that is obtained from natural salt deposits.

The products of the chloralkali industry have a number of important uses:

Chlorine is used:

- To purify water

- As a disinfectant
- In the production of:
    - Hypochlorous acid (used to kill bacteria in drinking water).
    - Paper, food.
    - Antiseptics, insecticides, medicines, textiles, laboratory chemicals.
    - Paints, petroleum products, solvents, plastics (such as polyvinyl chloride).

Sodium hydroxide (also known as 'caustic soda') is used to:

- Make soap and other cleaning agents.
- Purify bauxite (the ore of aluminium).
- Make paper.
- Make rayon (artificial silk).

One of the problems of producing chlorine and sodium hydroxide is that when they are produced together the chlorine combines with the sodium hydroxide to form chlorate (ClO^-) and chloride (Cl^-) ions. This leads to the production of sodium chlorate, NaClO, a component of household bleach.

To overcome this problem the chlorine and sodium hydroxide must be separated from each other so that they don't react. There are three industrial processes that have been designed to overcome this problem. All three methods involve electrolytic cells.

1. The Mercury Cell In the mercury-cell:
    - The anode is a carbon electrode suspended from the top of a chamber.
    - The cathode is liquid mercury that flows along the floor of this chamber.
    - The electrolyte is brine (NaCl solution) that is passed through the chamber.
    - When an electric current is applied to the circuit, chloride ions in the electrolyte are oxidised at the anode form chlorine gas.

      $2Cl^-(aq) \rightarrow Cl_2(g) + 2e^-$
    - At the same time sodium ions are reduced at the anode to solid sodium. The solid sodium dissolves in the mercury making a sodium/mercury amalgam.

      $Na^+(aq) + Hg(l) + e^- \rightarrow Na(Hg)$
    - The amalgam is poured into a separate vessel, where it decomposes into sodium and mercury.
    - The sodium reacts with water in the vessel and produces sodium hydroxide

and hydrogen gas, while the mercury returns to the electrolytic cell to be used again.

$2Na(Hg) + 2H_2O(l) \rightarrow 2NaOH(aq) + H_2(g) + Hg(l)$

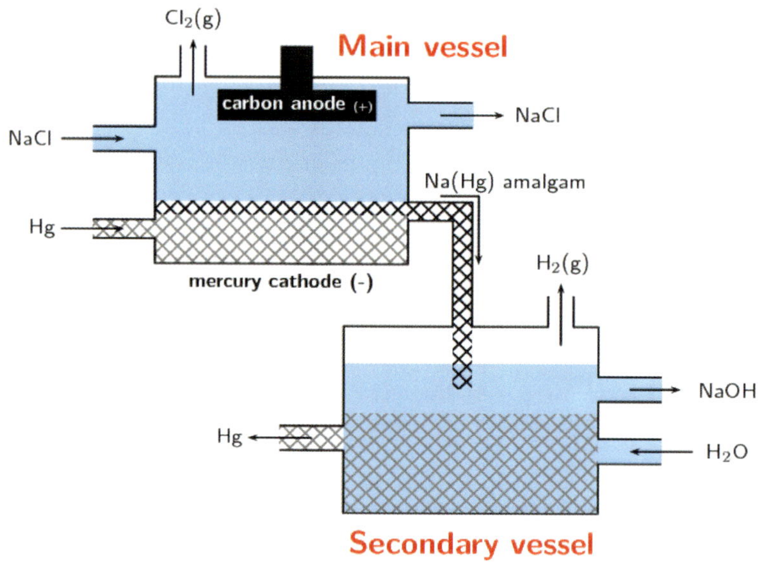

The mercury cell

The image above gives a good demonstration of how a mercury cell works.

This method only produces a fraction of the chlorine and sodium hydroxide that is used by industry as it has certain disadvantages:

- Mercury is expensive and toxic.
- Some mercury always escapes with the brine that has been used.
- Mercury reacts with the brine to form mercury(II) chloride.
- The mercury cell requires a lot of electricity.
- Although the chlorine gas produced is very pure, mercury has to be removed from the sodium hydroxide and hydrogen gas mixture.

In the past the effluent was released into lakes and rivers, causing mercury to accumulate in fish and other animals feeding on the fish. Today, the brine is treated before it is discharged so that the environmental impact is lower.

2. The Diaphragm Cell In the diaphragm-cell:

    - A porous diaphragm divides the electrolytic cell into an anode compartment and a cathode compartment
    - Brine is introduced into the anode compartment and flows through the diaphragm into the cathode compartment

- An electric current is passed through the brine causing the salt's chlorine ions and sodium ions to move to the electrodes
- Chlorine gas is produced at the anode $2Cl^- (aq) + 2e^- \rightarrow Cl_2(g)$
- At the cathode, sodium ions react with water forming caustic soda (NaOH) and hydrogen gas

  $2Na^+ (aq) + 2H_2O(l) + e^- \rightarrow 2NaOH(aq) + H_2(g)$
- Some NaCl salt remains in the solution with the caustic soda and can be removed at a later stage.

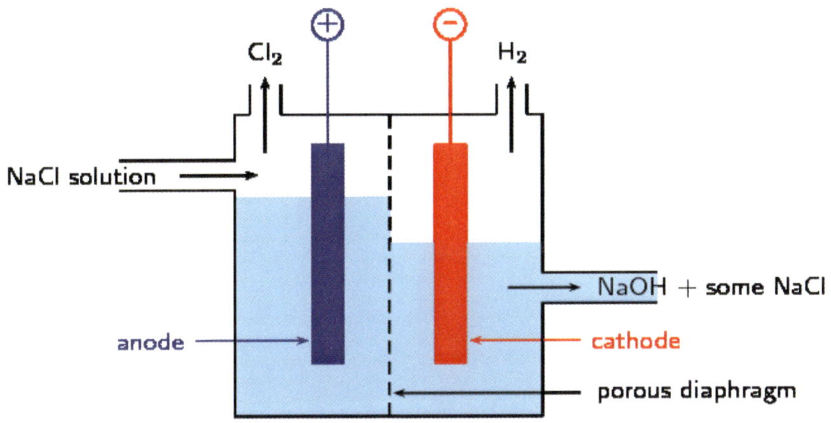

The diaphragm cell

The image above gives a good demonstration of how a diaphragm cell works.

The advantages of the diaphragm cell are:

- Uses less energy than the mercury cell.
- Does not contain toxic mercury.

It also has disadvantages however:

- The sodium hydroxide is much less concentrated and not as pure
- The chlorine gas often contains oxygen gas as well
- The process is less cost-effective as the sodium hydroxide solution needs to be concentrated and purified before it can be used.

3. The Membrane Cell

The membrane cell is very similar to the diaphragm cell, and the same reactions occur. The main differences are:

- The two electrodes are separated by an ion-selective membrane, rather than by a diaphragm.

- The membrane structure allows cations to pass through it between compartments of the cell but does not allow anions to pass through (this has nothing to do with the size of the pores, but rather with the charge on the ions).
- Brine is pumped into the anode compartment, and only the positively charged sodium ions pass into the cathode compartment, which contains pure water.

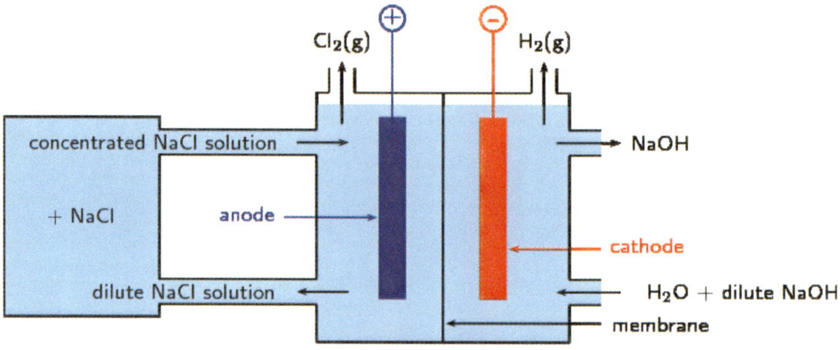

The membrane cell

The image above gives a good demonstration of how a membrane cell works.

- At the positively charged anode, Cl^- ions from the brine are oxidised to Cl_2 gas. $2Cl^-(aq) \rightarrow Cl_2(g) + 2e^-$
- At the negatively charged cathode, hydrogen ions in the water are reduced to hydrogen gas. $2H_2O(l) + 2e^- \rightarrow H_2(g) + 2OH^-$
- The Na^+ ions flow through the membrane to the cathode compartment and react with the remaining hydroxide (OH^-) ions from the water to form sodium hydroxide (NaOH).

  $Na^+ (aq) + OH^- (aq) \rightarrow NaOH(aq)$
- The chloride ions cannot pass through the membrane, so the chlorine does not come into contact with the sodium hydroxide in the cathode compartment. The sodium hydroxide is removed from the cell. The overall equation is as follows:

  $2NaCl(aq) + 2H_2O(l) \rightarrow Cl_2(g) + H_2(g) + 2NaOH(g)$

The advantages of using this method are:

- The sodium hydroxide that is produced is very pure because it is kept separate from the sodium chloride solution
- The sodium hydroxide has a relatively high concentration
- This process uses the least electricity of all three cells
- The cell is cheaper to operate than the other two cells
- The cell does not contain toxic mercury or asbestos.

Oxidation and Reduction

Oxidation-reduction reaction, also called redox reaction, any chemical reaction in which the oxidation number of a participating chemical species changes. The term covers a large and diverse body of processes. Many oxidation-reduction reactions are as common and familiar as fire, the rusting and dissolution of metals, the browning of fruit, and respiration and photosynthesis—basic life functions.

Gain and Loss of Oxygen

The earliest view of oxidation and reduction is that of adding oxygen to form an oxide (oxidation) or removing oxygen (reduction). They always occur together. For example, in the burning of hydrogen,

$$2H_2 + O_2 \rightarrow 2H_2O$$

the hydrogen is oxidized and the oxygen is reduced. The combination of nitrogen and oxygen which occurs at high temperatures follows the same pattern.

$$N_2 + O_2 \rightarrow 2NO$$

This formation of nitric oxide oxidizes the nitrogen and reduces the oxygen. In some reactions, the oxidation is most prominent. For example in the burning of methane,

$$CH_4 + 2O_2 \rightarrow CO_2 + 2H_2O$$

both carbon and hydrogen are oxidized (gain oxygen). The accompanying reduction of oxygen is perhaps easier to see when you describe reduction as the gaining of hydrogen.

On the other hand, the reaction of lead dioxide at high temperatures appears to be just reduction.

$$2PbO_2 \rightarrow 2PbO + O_2$$

The reduction of the lead dioxide is clear, but the associated oxidation of oxygen is easier to see when you describe oxidation as the losing of electrons.

Gain and Loss of Hydrogen

The original view of oxidation and reduction is that of adding or removing oxygen . An alternative approach is to describe oxidation as the loss of hydrogen and reduction as the gaining of hydrogen. This has an advantage in describing the burning of methane.

$$CH_4 + 2O_2 \rightarrow CO_2 + 2H_2O$$

With this approach it is clear that the carbon is oxidized (loses all four hydrogens) and that part of the oxygen is reduced (gains hydrogen). Another reaction where the hydrogen approach makes things clearer is the passing of methanol over a hot copper gauze to form formaldehyde and hydrogen gas:

$$CH_3OH \text{ -> } CH_2O + H_2$$

Both carbon-containing molecules have the same oxygen content, but the formation of the formaldehyde is seen to be oxidation because hydrogens are lost. The formation of H_2 is a reduction process as the two released hydrogens get together.

The formation of methanol from reacting carbon monoxide with hydrogen combines oxidation and reduction in the single molecular product:

$$CO + H_2 \text{ -> } CH_3OH$$

The CO is reduced because it gains hydrogen, and the hydrogen is oxidized by its association with the oxygen.

Gain and Loss of Electrons

The original view of oxidation and reduction is that of adding or removing oxygen. An alternative view is to describe oxidation as the losing of electrons and reduction as the gaining of electrons. One example in which this approach is of value is in the high temperature reaction of lead dioxide.

$$2PbO_2 \text{ -> } 2PbO + O_2$$

In this reaction the lead atoms gain an electron (reduction) while the oxygen loses electrons (oxidation).

This electron view of oxidation and reduction helps you deal with the fact that "oxidation" can occur even when there is no oxygen. The definition of redox reactions is extended to include other reactions with nonmetals such as chlorine and bromine. For example, the reaction,

$$Mg + Cl_2 \text{ -> } Mg^{2+} + 2Cl^-$$

Magnesium loses electrons and is therefore said to be "oxidized", whereas the chlorines gain electrons and are said to be reduced. Another way to judge that the chlorine has been reduced is the fact that the charge on the atoms is made more negative, or reduced. Treating that charge as an "oxidation number" is another way to characterize oxidation and reduction.

The view of oxidation and reduction as the loss and gain of electrons, respectively, is particularly appropriate for discussing reactions in electrochemical cells. For example, in the zinc-copper cell, the oxidation and reduction half-reactions are:

$Zn(s) \rightarrow Zn^{2+}(aq) + 2e^-$	The zinc "half-reaction" is classified as oxidation since it loses electrons. The terminal at which oxidation occurs is called the "anode". For a battery, this is the negative terminal.

The copper "half-reaction" is classified as reduction since it gains electrons. The terminal at which reduction occurs is called the "cathode". For a battery, this is the positive terminal.	$Cu^{2+}(aq) + 2e^- \rightarrow Cu(s)$

Oxidation Number

The original view of oxidation and reduction is that of adding or removing oxygen. An alternative view which is useful in dealing with ions is to define an oxidation number which is equal to the net charge of the product of a reaction. Oxidation is then viewed as a reaction which increases the oxidation number and reduction as one which reduces the oxidation number. This view of oxidation and reduction helps you deal with the fact that "oxidation" can occur even when there is no oxygen! The definition of redox reactions is extended to include other reactions with nonmetals such as chlorine and bromine. For example, in the reaction,

$$Mg + Cl_2 \rightarrow Mg^{2+} + 2Cl^-$$

the Mg is seen to increase in oxidation number from 0 to 2 (oxidation) while the chlorine atoms experience a decrease in oxidation number from 0 to -1 (reduction).

This is essentially equivalent to viewing oxidation and reduction as losing or gaining electrons, but may be easier to remember.

Oxidation Numbers versus the True Charge on Ions

The terms *ionic* and *covalent* describe the extremes of a continuum of bonding. There is some covalent character in even the most ionic compounds and vice versa.

It is useful to think about the compounds of the main group metals as if they contained positive and negative ions. The chemistry of magnesium oxide, for example, is easy to understand if we assume that MgO contains Mg^{2+} and O^{2-} ions. But no compounds are 100% ionic. There is experimental evidence, for example, that the true charge on the magnesium and oxygen atoms in MgO is +1.5 and -1.5.

Oxidation states provide a compromise between a powerful model of oxidation-reduction reactions based on the assumption that these compounds contain ions and our knowledge that the true charge on the ions in these compounds is not as large as this model predicts. By definition, the oxidation state of an atom is the charge that atom would carry if the compound were purely ionic.

For the active metals in Groups IA and IIA, the difference between the oxidation state of the metal atom and the charge on this atom is small enough to be ignored.

The main group metals in Groups IIIA and IVA, however, form compounds that have a significant amount of covalent character. It is misleading, for example, to assume that aluminum bromide contains Al^{3+} and Br^- ions. It actually exists as Al_2Br_6 molecules.

This problem becomes even more severe when we turn to the chemistry of the transition metals. MnO, for example, is ionic enough to be considered a salt that contains Mn^{2+} and O^{2-} ions. Mn_2O_7, on the other hand, is a covalent compound that boils at room temperature. It is therefore more useful to think about this compound as if it contained manganese in a +7 oxidation state, not Mn^{7+} ions.

Oxidizing and Reducing Agents

In redox processes, the reductant transfers electrons to the oxidant. Thus, in the reaction, the reductant or *reducing agent* loses electrons and is oxidized, and the oxidant or *oxidizing agent* gains electrons and is reduced. The pair of an oxidizing and reducing agent that are involved in a particular reaction is called a redox pair. A redox couple is a reducing species and its corresponding oxidizing form, e.g., Fe^{2+}/Fe^{3+}.

Oxidizers

The international pictogram for oxidizing chemicals

Substances that have the ability to oxidize other substances (cause them to lose electrons) are said to be oxidative or oxidizing and are known as oxidizing agents, oxidants, or oxidizers. That is, the oxidant (oxidizing agent) removes electrons from another substance, and is thus itself reduced. And, because it "accepts" electrons, the oxidizing agent is also called an electron acceptor. Oxygen is the quintessential oxidizer.

Oxidants are usually chemical substances with elements in high oxidation states (e.g., H_2O_2 , MnO_4^- , CrO_3 , $Cr_2O_7^{2-}$, OsO_4), or else highly electronegative elements (O_2, F_2, Cl_2, Br_2) that can gain extra electrons by oxidizing another substance.

Reducers

Substances that have the ability to reduce other substances (cause them to gain electrons) are said to be reductive or reducing and are known as reducing agents, reductants, or reducers. The reductant (reducing agent) transfers electrons to another substance, and is thus itself oxidized. And, because it "donates" electrons, the reducing agent is also called an electron donor. Electron donors can also form charge transfer complexes with electron acceptors.

Reductants in chemistry are very diverse. Electropositive elemental metals, such as lithium, sodium, magnesium, iron, zinc, and aluminium, are good reducing agents. These metals donate or *give away* electrons readily. *Hydride transfer reagents*, such as $NaBH_4$ and $LiAlH_4$, are widely used in organic chemistry, primarily in the reduction of carbonyl compounds to alcohols. Another method of reduction involves the use of hydrogen gas (H_2) with a palladium, platinum, or nickel catalyst. These *catalytic reductions* are used primarily in the reduction of carbon-carbon double or triple bonds.

Standard Electrode Potentials (Reduction Potentials)

Each half-reaction has a *standard electrode potential* (E^o_{cell}), which is equal to the potential difference or voltage at equilibrium under standard conditions of an electrochemical cell in which the cathode reaction is the half-reaction considered, and the anode is a standard hydrogen electrode where hydrogen is oxidized:

$$\tfrac{1}{2}\,H_2 \rightarrow H^+ + e^-.$$

The electrode potential of each half-reaction is also known as its *reduction potential* E^o_{red}, or potential when the half-reaction takes place at a cathode. The reduction potential is a measure of the tendency of the oxidizing agent to be reduced. Its value is zero for $H^+ + e^- \rightarrow \tfrac{1}{2}\,H_2$ by definition, positive for oxidizing agents stronger than H^+ (e.g., +2.866 V for F_2) and negative for oxidizing agents that are weaker than H^+ (e.g., −0.763 V for Zn^{2+}).

For a redox reaction that takes place in a cell, the potential difference is:

$$E^o_{cell} = E^o_{cathode} - E^o_{anode}$$

However, the potential of the reaction at the anode was sometimes expressed as an *oxidation potential*:

$$E^o_{ox} = -E^o_{red}.$$

The oxidation potential is a measure of the tendency of the reducing agent to be oxidized, but does not represent the physical potential at an electrode. With this notation, the cell voltage equation is written with a plus sign,

$$E^o_{cell} = E^o_{red(cathode)} + E^o_{ox(anode)}$$

Examples of Redox Reactions

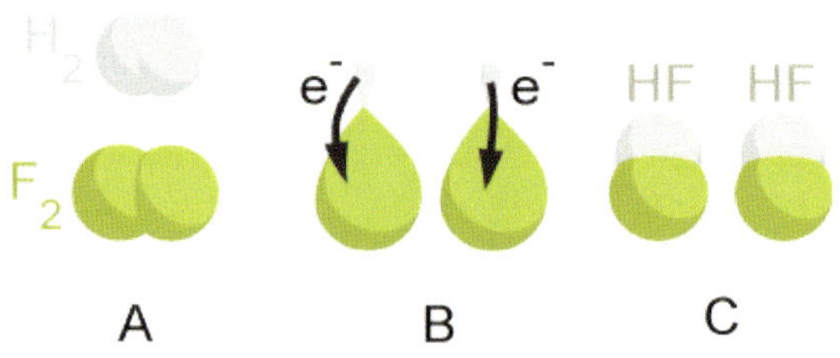

Illustration of a redox reaction

A good example is the reaction between hydrogen and fluorine in which hydrogen is being oxidized and fluorine is being reduced:

$$H_2 + F_2 \rightarrow 2\ HF$$

We can write this overall reaction as two half-reactions:

- The oxidation reaction:

  $$H_2 \rightarrow 2\ H^+ + 2\ e^-$$

- The reduction reaction:

  $$F_2 + 2\ e^- \rightarrow 2\ F^-$$

Analyzing each half-reaction in isolation can often make the overall chemical process clearer. Because there is no net change in charge during a redox reaction, the number of electrons in excess in the oxidation reaction must equal the number consumed by the reduction reaction.

Elements, even in molecular form, always have an oxidation state of zero. In the first half-reaction, hydrogen is oxidized from an oxidation state of zero to an oxidation state of +1. In the second half-reaction, fluorine is reduced from an oxidation state of zero to an oxidation state of −1.

When adding the reactions together the electrons are canceled:

H_2	→	$2\ H^+ + 2\ e^-$
$F_2 + 2\ e^-$	→	$2\ F^-$
$H_2 + F_2$	→	$2\ H^+ + 2\ F^-$

And the ions combine to form hydrogen fluoride:

$$2\ H^+ + 2\ F^- \rightarrow 2\ HF$$

The overall reaction is:

$$H_2 + F \rightarrow 2\ HF$$

Metal Displacement

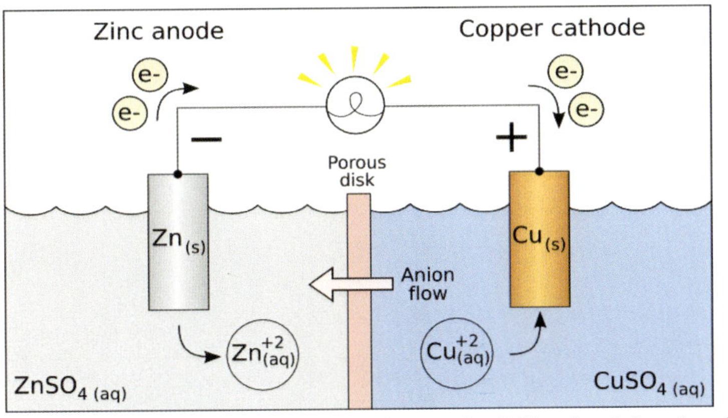

A redox reaction is the force behind an electrochemical cell like the Galvanic cell pictured. The battery is made out of a zinc electrode in a $ZnSO_4$ solution connected with a wire and a porous disk to a copper electrode in a $CuSO_4$ solution

In this type of reaction, a metal atom in a compound (or in a solution) is replaced by an atom of another metal. For example, copper is deposited when zinc metal is placed in a copper(II) sulfate solution:

$$Zn(s) + CuSO_4(aq) \rightarrow ZnSO_4(aq) + Cu(s)$$

In the above reaction, zinc metal displaces the copper(II) ion from copper sulfate solution and thus liberates free copper metal.

The ionic equation for this reaction is:

$$Zn + Cu^{2+} \rightarrow Zn^{2+} + Cu$$

As two half-reactions, it is seen that the zinc is oxidized:

$$Zn \rightarrow Zn^{2+} + 2\ e^-$$

And the copper is reduced:

$$Cu^{2+} + 2\ e^- \rightarrow Cu$$

Other Examples

- The reduction of nitrate to nitrogen in the presence of an acid (denitrification):

  $$2\ NO_3^- + 10\ e^- + 12\ H^+ \rightarrow N_2 + 6\ H_2O$$

- The combustion of hydrocarbons, such as in an internal combustion engine, which produces water, carbon dioxide, some partially oxidized forms such as carbon monoxide, and heat energy. Complete oxidation of materials containing carbon produces carbon dioxide.

- In organic chemistry, the stepwise oxidation of a hydrocarbon by oxygen produces water and, successively, an alcohol, an aldehyde or a ketone, a carboxylic acid, and then a peroxide.

Corrosion and Rusting

Oxides, such as iron(III) oxide or rust, which consists of hydrated iron(III) oxides $Fe_2O_3 \cdot nH_2O$ and iron(III) oxide-hydroxide ($FeO(OH)$, $Fe(OH)_3$), form when oxygen combines with other elements

Iron rusting in pyrite cubes

- The term corrosion refers to the electrochemical oxidation of metals in reaction with an oxidant such as oxygen. Rusting, the formation of iron oxides, is a well-known example of electrochemical corrosion; it forms as a result of the oxidation of iron metal. Common rust often refers to iron(III) oxide, formed in the following chemical reaction:

  $4\ Fe + 3\ O_2 \rightarrow 2\ Fe_2O_3$

- The oxidation of iron(II) to iron(III) by hydrogen peroxide in the presence of an acid:

  $Fe^{2+} \rightarrow Fe^{3+} + e^-$

  $H_2O_2 + 2\ e^- \rightarrow 2\ OH^-$

Overall equation:

$$2\ Fe^{2+} + H_2O_2 + 2\ H^+ \rightarrow 2\ Fe^{3+} + 2\ H_2O$$

Redox Reactions in Industry

Cathodic protection is a technique used to control the corrosion of a metal surface by making it the cathode of an electrochemical cell. A simple method of protection connects protected metal to a more easily corroded "sacrificial anode" to act as the anode. The sacrificial metal instead of the protected metal, then, corrodes. A common application of cathodic protection is in galvanized steel, in which a sacrificial coating of zinc on steel parts protects them from rust.

Oxidation is used in a wide variety of industries such as in the production of cleaning products and oxidizing ammonia to produce nitric acid, which is used in most fertilizers.

Redox reactions are the foundation of electrochemical cells, which can generate electrical energy or support electrosynthesis. Metal ores often contain metals in oxidized states such as oxides or sulfides, from which the pure metals are extracted by smelting at high temperature in the presence of a reducing agent. The process of electroplating uses redox reactions to coat objects with a thin layer of a material, as in chrome-plated automotive parts, silver plating cutlery, galvanization and gold-plated jewelry.

Redox Reactions in Biology

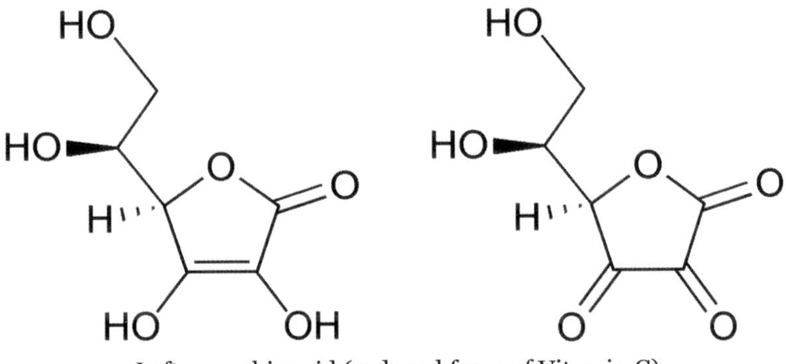

Left: ascorbic acid (reduced form of Vitamin C)
Right: dehydroascorbic acid (oxidized form of Vitamin C)

Many important biological processes involve redox reactions.

Cellular respiration, for instance, is the oxidation of glucose ($C_6H_{12}O_6$) to CO_2 and the reduction of oxygen to water. The summary equation for cell respiration is:

$$C_6H_{12}O_6 + 6\ O_2 \rightarrow 6\ CO_2 + 6\ H_2O$$

The process of cell respiration also depends heavily on the reduction of NAD^+ to NADH and the reverse reaction (the oxidation of NADH to NAD^+). Photosynthesis and cellular respiration are complementary, but photosynthesis is not the reverse of the redox reaction in cell respiration:

$$6\ CO_2 + 6\ H_2O + \text{light energy} \rightarrow C_6H_{12}O_6 + 6\ O_2$$

Biological energy is frequently stored and released by means of redox reactions. Photosynthesis involves the reduction of carbon dioxide into sugars and the oxidation of water into molecular oxygen. The reverse reaction, respiration, oxidizes sugars to produce carbon dioxide and water. As intermediate steps, the reduced carbon compounds are used to reduce nicotinamide adenine dinucleotide (NAD^+) to NADH, which then contributes to the creation of a proton gradient, which drives the synthesis of adenosine triphosphate (ATP) and is maintained by the reduction of oxygen. In animal cells, mitochondria perform similar functions.

Free radical reactions are redox reactions that occur as a part of homeostasis and killing microorganisms, where an electron detaches from a molecule and then reattaches almost instantaneously. Free radicals are a part of redox molecules and can become harmful to the human body if they do not reattach to the redox molecule or an antioxidant. Unsatisfied free radicals can spur the mutation of cells they encounter and are, thus, causes of cancer.

The term redox state is often used to describe the balance of GSH/GSSG, NAD^+/NADH and $NADP^+$/NADPH in a biological system such as a cell or organ. The redox state is reflected in the balance of several sets of metabolites (e.g., lactate and pyruvate, beta-hydroxybutyrate, and acetoacetate), whose interconversion is dependent on these ratios. An abnormal redox state can develop in a variety of deleterious situations, such as hypoxia, shock, and sepsis. Redox mechanism also control some cellular processes. Redox proteins and their genes must be co-located for redox regulation according to the CoRR hypothesis for the function of DNA in mitochondria and chloroplasts.

Redox Cycling

A wide variety of aromatic compounds are enzymatically reduced to form free radicals that contain one more electron than their parent compounds. In general, the electron donor is any of a wide variety of flavoenzymes and their coenzymes. Once formed, these anion free radicals reduce molecular oxygen to superoxide, and regenerate the unchanged parent compound. The net reaction is the oxidation of the flavoenzyme's coenzymes and the reduction of molecular oxygen to form superoxide. This catalytic behavior has been described as futile cycle or redox cycling.

Redox Reactions in Geology

In geology, redox is important to both the formation of minerals and the mobilization of minerals, and is also important in some depositional environments. In general, the redox state of most rocks can be seen in the color of the rock. The rock forms in oxidizing conditions, giving it a red color. It is then "bleached" to a green—or sometimes white—form when a reducing fluid passes through the rock. The reduced fluid can also carry uranium-bearing minerals. Famous examples of redox conditions affecting geological processes include uranium deposits and Moqui marbles.

Mi Vida uranium mine, near Moab, Utah. The alternating red and white/green bands of sandstone correspond to oxidized and reduced conditions in groundwater redox chemistry

Balancing Redox Reactions

Describing the overall electrochemical reaction for a redox process requires a *balancing* of the component half-reactions for oxidation and reduction. In general, for reactions in aqueous solution, this involves adding H^+, OH^-, H_2O, and electrons to compensate for the oxidation changes.

Acidic Media

In acidic media, H^+ ions and water are added to half-reactions to balance the overall reaction.

For instance, when manganese(II) reacts with sodium bismuthate:

Unbalanced reaction:	$Mn^{2+}(aq) + NaBiO_3(s) \rightarrow Bi^{3+}(aq) + MnO_4^-(aq)$
Oxidation:	$4\ H_2O(l) + Mn^{2+}(aq) \rightarrow MnO_4^-(aq) + 8\ H^+(aq) + 5\ e^-$
Reduction:	$2\ e^- + 6\ H^+ + BiO_3^-(s) \rightarrow Bi^{3+}(aq) + 3\ H_2O(l)$

The reaction is balanced by scaling the two half-cell reactions to involve the same number of electrons (multiplying the oxidation reaction by the number of electrons in the reduction step and vice versa):

$$8\ H_2O(l) + 2\ Mn^{2+}(aq) \rightarrow 2\ MnO_4^-(aq) + 16\ H^+(aq) + 10\ e^-$$

$$10\ e^- + 30\ H^+ + 5\ BiO_3^-(s) \rightarrow 5\ Bi^{3+}(aq) + 15\ H_2O(l)$$

Adding these two reactions eliminates the electrons terms and yields the balanced reaction:

$$14\ H^+(aq) + 2\ Mn^{2+}(aq) + 5\ NaBiO_3(s) \rightarrow 7\ H_2O(l) + 2\ MnO_4^-\ (aq) + 5\ Bi^{3+}(aq) + 5\ Na^+(aq)$$

Basic Media

In basic media, OH^- ions and water are added to half reactions to balance the overall reaction.

For example, in the reaction between potassium permanganate and sodium sulfite:

Unbalanced reaction:	$KMnO_4 + Na_2SO_3 + H_2O \rightarrow MnO_2 + Na_2SO_4 + KOH$
Reduction:	$3\ e^- + 2\ H_2O + MnO_4^- \rightarrow MnO_2 + 4\ OH^-$
Oxidation:	$2\ OH^- + SO_3^{2-} \rightarrow SO_4^{2-} + H_2O + 2\ e^-$

Balancing the number of electrons in the two half-cell reactions gives:

$$6\ e^- + 4\ H_2O + 2\ MnO^-\backslash_4 \rightarrow 2\ MnO_2 + 8\ OH^-$$

$$6\ OH^- + 3\ SO_3^{2-} \rightarrow 3\ SO_4^{2-} + 3\ H_2O + 6\ e^-$$

Adding these two half-cell reactions together gives the balanced equation:

$$2\ KMnO_4 + 3\ Na_2SO_3 + H_2O \rightarrow 2\ MnO_2 + 3\ Na_2SO_4 + 2\ KOH$$

Electrochemical Cell

An electrochemical cell is a device that produces an electric current from energy released by a spontaneous redox reaction. This kind of cell includes the galvanic, or voltaic, cell, named after Luigi Galvani and Alessandro Volta. These scientists conducted several experiments on chemical reactions and electric current during the late 18th century.

Electrochemical cells have two conductive electrodes, called the anode and the cathode. The anode is defined as the electrode where oxidation occurs. The cathode is the electrode where reduction takes place. Electrodes can be made from any sufficiently conductive materials, such as metals, semiconductors, graphite, and even conductive polymers. In between these electrodes is the electrolyte, which contains ions that can freely move.

Types of Cell

There are two fundamental types of electrochemical cell: galvanic and electrolytic.

Galvanic cells convert chemical potential energy into electrical energy.

The energy conversion is achieved by spontaneous ($\Delta G < 0$) redox reactions producing a flow of electrons.

Electrolytic cells are driven by an external source of electrical energy.

A flow of electrons drives non-spontaneous ($\Delta G \geq 0$) redox reactions.

Examples

- A battery powering something is an example of a galvanic cell.
- Rechargeable batteries are examples of both types of cell: they operate as galvanic cells when they are powering a device and as electrolytic cells when they are being recharged.
- Examples of electrolytic cells also include those used to split water into hydrogen and oxygen, and those that convert aluminum ore to aluminum metal.

Electrochemical Cell Diagram

Let us explain the working of an electrochemical cell with the help of the following diagram.

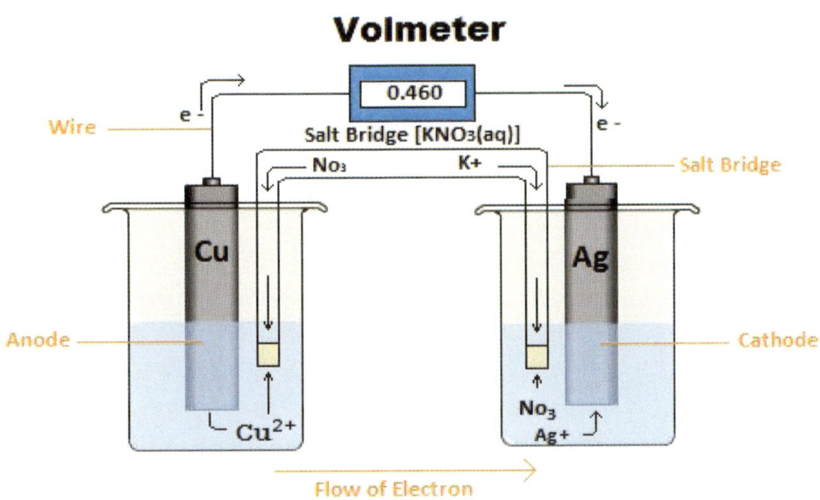

The arrangement consists of two beakers, one of which contains $CuSO_4$ solution and the other one contains $AgNO_3$ solution. A copper rod is dipped in the $CuSO_4$ solution and a silver rod is dipped in the $AgNO_3$ solution. These metallic rods are called Electrodes. The solutions in the two beakers are connected by an inverted U tube containing a saturated solution of KNO_3, which does not undergo any chemical change during the process. This U tube is called a Salt Bridge. A volt meter is attached between the two electrodes to measure the potential difference between the two electrodes.

When the circuit is completed, it is observed that an electric current flows through the external circuit, which we can detect by placing an ammeter in the circuit.

These observations can be explained as follows:

During the reaction, Cu is oxidized to Cu^{2+} ions, which go into the solution.

$$Cu(s) \rightarrow Cu^{2+}(aq) + 2e^{-} \text{ (OXIDATION)}$$

Therefore, the copper rod gradually loses its weight. The electrons released at the electrodes move towards the other electrode through the outer electrode. These are accepted by Ag+ ions of $AgNO_3$ solution which are reduced to silver metal and it gets deposited on the silver electrode.

$$2Ag^{+}(aq) + 2e^{-} \rightarrow 2Ag(s) \text{ (REDUCTION)}$$

Important Functions of Salt Bridge

- Salt bridge completes the electrical circuit.
- Salt bridge maintains the electrical neutrality of the two half cell solutions.

Electrochemical Cell Notation

An electrochemical cell consists of two electrodes Anode and Cathode. The electrolyte solution containing these electrodes are called Half cells. When the two half cells are combined, a Cell is formed. The following conventions are used to represent an electrochemical cell.

1. A single vertical line will indicate a change in state or phase.
2. Inside the half-cell, the reactants are listed before the products.
3. Aqueous solutions concentrations are written in parentheses after the symbol for the ion or molecule.
4. Double vertical line would be used to indicate the junction between the half-cells or the salt bridge.
5. The line notation for the anode (oxidation) is written before the line notation for the cathode (reduction).

Example:- 1

The following cell reaction is given

$$Zn(s) + Cu^{2+}(aq) \rightarrow Cu(s) + Zn^{2+}(aq)$$

Cell Notation:- $Zn \mid Zn^{2+}(1.0\ M) \mid\mid Cu^{2+}(1.0\ M) \mid Cu$

Example:- 2

Ni- $AgNO_3$ cell

Oxidation half reaction $Ni(s) \rightarrow Ni^{2+}(aq) + 2e^-$

Reduction half reaction $2Ag^+(aq) + 2e^- \rightarrow 2Ag(s)$

Overall cell reaction $Ni(s) + 2Ag^+(aq) \rightarrow 2Ag(s) + Ni^{2+}(aq)$

Cell Notation :- $Ni(s) \mid Ni^{2+}(aq) \mid\mid Ag^+(aq) \mid Ag(s)$

Example:- 3

Zn – HCl reaction

Oxidation half reaction $Zn(s) \rightarrow Zn^{2+}(aq) + 2e^-$

Reduction half reaction $2H^+ + 2e^- \rightarrow H_2(g)$

Overall cell reaction $Zn(s) + 2H^+(aq) \rightarrow Zn^{2+}(aq) + H_2(g)$

Cell Notation :- $Zn(s) \mid Zn^{2+}(aq) \mid\mid H^+(aq) \mid H_2(g)$

Working of Electrochemical Cell

When the circuit is completed, a deflection is observed in the (G) towards the zinc electrode indicating that the e^- are flowing from the Zn electrode to Cu electrode.

At the Zn electrode, oxidation takes place.

$$Zn \rightarrow Zn^{+2} + 2e^- \text{ (oxidation)} \quad (1)$$

The electron 'e' is removed, lost or retained by the metal move through the material contained in the electrode, and reach the Copper electrode at which they are accepted by Copper ions of the solution to form the neutral Copper atoms.

$$Cu^{+2} + 2e^- \rightarrow Cu \qquad (2)$$

In an electrochemical cell, each electrode constitutes one half of the cell and the reaction taking place at the electrode is called half-cell reaction. The overall cell reaction is obtained by adding the two half-cell reactions (1) and (2).

$$Zn + Cu^{+2} \rightarrow Zn^{+2} + Cu \text{ (Overall Reaction)}$$

From this, it is found that, when Zn is added to $CuSO_4$ solution, Zn displaces Cu from $CuSO_4$ with the liberation of heat. But, in the electrochemical cell, there is no direct contact between Zn and $CuSO_4$. The heat energy that would have been liberated, appears in the form of electrical energy. Hence, the electrochemical cell acts as a source of current, although for a short interval.

Accordingly, in the above constructed electrochemical cell, zinc electrode acts as negative electrode while Cu electrode acts as the positive electrode.

In general, KNO_3 salt bridge is used when silver electrode is involved as one of the electrodes.

Electrochemical Cell Equation

In general, the electrode at which reduction takes place is written on the RHS of the salt bridge and the electrode at which oxidation takes place is written on the LHS of the salt bridge. The salt bridge linking the aqueous solutions is represented by two vertical parallel lines having ions on both sides.

Left hand side	**Right hand side**
Anode	Cathode
(Oxidation)	(Reduction)

In the above system, Zn electrode is written on the LHS while the Cu electrode is written on the RHS of the salt bridge.

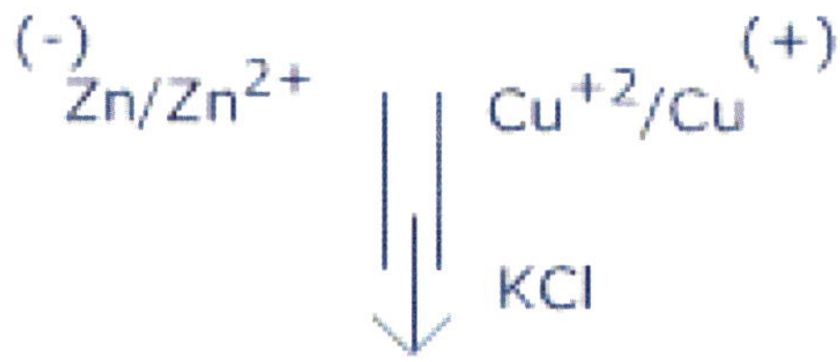

In general, of the two electrodes, the electrode having higher SRP (standard reduction potential) acts as the positive electrode.

Example: Zn| Zn^{2+} (C_1) | | Cu^{2+} (C_2)| Cu

Metal/Metal ion (Conc.) || Metal ion (Conc.) | Metal,

C_1 and C_2 are the concentrations of Zinc and Copper ions respectively.

||-Represents the salt bridge.

Electrochemical Cell Examples

Below you could see examples.

Solved Example:

Question: Depict the Galvanic cell in which the following reaction takes place

$$Zn(s) + 2Ag^+(aq) \rightarrow Zn^{2+}(aq) + 2Ag(s)$$

Also show individual reactions at each electrode.

Solution:

The cell notation is $Zn(s) \mid Zn^{2+}(aq) \mid\mid Ag^{+}(aq) \mid Ag(s)$

Reaction at Anode: $Zn(s) \rightarrow Zn^{2+}(aq) + 2e^{-}$

Reaction at cathode: $2Ag^{+}(aq) + 2e^{-} \rightarrow 2Ag(s)$

Galvanic Cell

The Galvanic Cells is a device that transforms chemical energy into electric energy. It is the common application of electro chemistry which is also called battery. It was invented by *Luigi Galvani and Alessandro Volta* which has capability of producing a voltage.

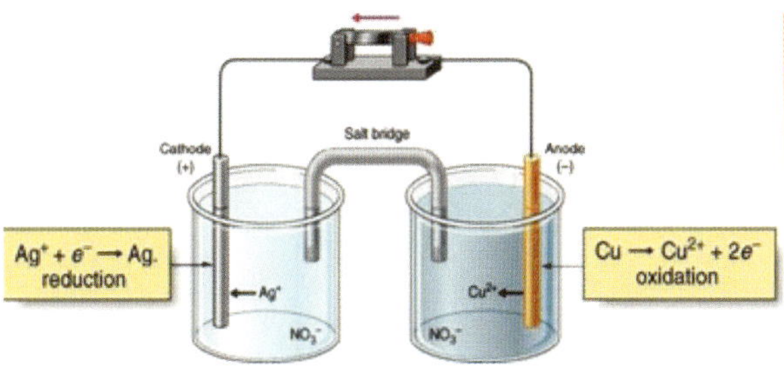

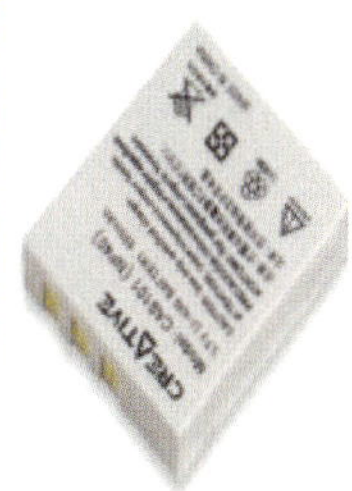

In this cell there is a container in which a solution of concentrated Copper Sulphate ($CuSO_4$) is kept inside it and a copper rod is inserted inside the solution of $CuSO_4$ which act like cathode. Inside this container a porous container is kept in which concentrated Sulphuric Acid (H_2SO_4) is filled in which a zinc rod is inserted in it which acts like anode. Thus when a wire is connected through copper rod and zinc rod an electric current starts to flow.

Important Components of Galvanic Cells

- The anode is an electrode where oxidation occurs.
- The cathode is an electrode where reduction occurs.
- A salt bridge is a chamber of electrolytes necessary to complete the circuit in a voltaic cell.
- The oxidation and reduction reactions are separated into compartments called half-cells.
- The external circuit is used to conduct the flow of electrons between the electrodes of the voltaic cell and usually includes a load.
- The load is the part of the circuit which utilizes the flow of electrons to perform some function.

Working of Galvanic Cells

The Galvanic cells have two conductive electrodes, called the anode and the cathode. The anode is an electrode where oxidation occurs. The cathode is an electrode where reduction takes place. Electrodes can be made from any sufficiently conductive materials, such as metals, semiconductors, graphite, and even conductive polymers. In between these electrodes is the electrolyte, which contains ions that can freely move.

The cell uses two different metal electrodes, each in an electrolyte solution. The anode will undergo oxidation and the cathode will undergo reduction. The metal of the anode will oxidize, going from an oxidation state of 0 (in the solid form) to a positive oxidation state, and it will become an ion. At the cathode, the metal ion in the solution will accept one or more electrons from the cathode, and the ion's oxidation state will reduce to 0. This forms a solid metal that deposits on the cathode. The two electrodes must be electrically connected to each other, allowing for a flow of electrons that leave the metal of the anode and flow through this connection to the ions at the surface of the cathode. This flow of electrons is an electrical current that can be used to do work, such as turn a motor or power a light.

Cell Voltage

The standard electrical potential of a cell can be determined by the use of a standard potential table for the two half cells involved. The first step is to identify the two metals reacting in the cell. Then one looks up the standard electrode potential, E^o, in volts, for each of the two half reactions. The standard potential for the cell is equal to the more positive E^o value minus the more negative E^o value.

For example, solutions of $CuSO_4$ and $ZnSO_4$, each solution has a corresponding metal strip in it, and a salt bridge or porous disk connecting the two solutions and allowing SO_4^{2-} ions to flow freely between the copper and zinc solutions. To calculate the standard potential one looks up copper and zinc's half reactions and finds:

$$Cu^{2+} + 2\,e^- \rightleftharpoons Cu: E^o = +0.34 \text{ V}$$

$$Zn^{2+} + 2\,e^- \rightleftharpoons Zn: E^o = -0.76 \text{ V}$$

Thus the overall reaction is:

$$Cu^{2+} + Zn \rightleftharpoons Cu + Zn^{2+}$$

The standard potential for the reaction is then +0.34 V – (−0.76 V) = 1.10 V. The polarity of the cell is determined as follows. Zinc metal is more strongly reducing than copper metal; equivalently, the standard (reduction) potential for zinc is more negative than that of copper. Thus, zinc metal will lose electrons to copper ions and develop a positive electrical charge. The equilibrium constant, K, for the cell is given by,

$$\ln K = \frac{nFE^{\mathrm{o}}}{RT}$$

where F is the Faraday constant, R is the gas constant and T is the temperature in kelvins. For the Daniell cell K is approximately equal to 1.5×10^{37}. Thus, at equilibrium, a few electrons are transferred, enough to cause the electrodes to be charged.

Actual half-cell potentials must be calculated by using the Nernst equation as the solutes are unlikely to be in their standard states,

$$E_{\text{half-cell}} = E^{\mathrm{o}} - \frac{RT}{nF}\ln_e Q$$

where Q is the reaction quotient. This simplifies to,

$$E_{\text{half-cell}} = E^{\mathrm{o}} - 2.303\frac{RT}{nF}\log_{10}\{M^{n+}\}$$

where $\{M^{n+}\}$ is the activity of the metal ion in solution. The metal electrode is in its standard state so by definition has unit activity. In practice concentration is used in place of activity. The potential of the whole cell is obtained by combining the potentials for the two half-cells, so it depends on the concentrations of both dissolved metal ions.

The value of $2.303R/F$ is 0.19845×10^{-3} V/K, so at 25 °C (298.15 K) the half-cell potential will change by if the concentration of a metal ion is increased or decreased by a factor of 10.

$$E_{\text{half-cell}} = E^{\mathrm{o}} - \frac{0.05918\ \mathrm{V}}{n}\log_{10}[M^{n+}]$$

These calculations are based on the assumption that all chemical reactions are in equilibrium. When a current flows in the circuit, equilibrium conditions are not achieved and the cell potential will usually be reduced by various mechanisms, such as the development of overpotentials. Also, since chemical reactions occur when the cell is producing power, the electrolyte concentrations change and the cell voltage is reduced. A consequence of the temperature dependency of standard potentials is that the voltage produced by a galvanic cell is also temperature dependent.

Galvanic Corrosion

Galvanic corrosion is a process that degrades metals electrochemically. This corrosion occurs when two dissimilar metals are located in contact with each other in the presence of an electrolyte, such as salt water, forming a galvanic cell. A cell can also be formed if the same metal is exposed to two different concentrations of electrolyte. The resulting electrochemical potential then develops an electric current that electrolytically dissolves the less noble material.

Electrolytic Cell

Electrolytic cell, any device in which electrical energy is converted to chemical energy, or vice versa. Such a cell typically consists of two metallic or electronic conductors (electrodes) held apart from each other and in contact with an electrolyte (q.v.), usually a dissolved or fused ionic compound. Connection of the electrodes to a source of direct electric current renders one of them negatively charged and the other positively charged. Positive ions in the electrolyte migrate to the negative electrode (cathode) and there combine with one or more electrons, losing part or all of their charge and becoming new ions having lower charge or neutral atoms or molecules; at the same time, negative ions migrate to the positive electrode (anode) and transfer one or more electrons to it, also becoming new ions or neutral particles. The overall effect of the two processes is the transfer of electrons from the negative ions to the positive ions, a chemical reaction. An example is the electrolysis of sodium chloride (common salt), forming sodium metal and chlorine gas; the energy required to make the reaction proceed is supplied by the electric current. Other common applications of electrolysis include electrodeposition for refining or plating of metals and the production of caustic soda.

In the case of substances that generate energy, rather than consume it, when they react with each other, some or all of this energy can be converted to electricity if the reaction can be divided into an oxidation and a reduction that can be made to occur at separate electrodes. In the lead-acid storage battery, for example, lead dioxide, lead metal, and sulfuric acid react to form lead sulfate and water; the separate processes are the oxidation of lead to lead sulfate at one electrode and the reduction of lead dioxide to lead sulfate at the other while electric charge is transported through the electrolyte by the migration of hydrogen ions. These processes create a driving force (a voltage or electrical potential) that causes electricity to flow through an external circuit joining the two electrodes. Many other chemical combinations have been utilized in cells and batteries.

Other cells for generating electricity by means other than motion of a conductor in a magnetic field include solar cells, in which electron flow between semiconductors results from absorption of light, and fuel cells, in which a continuous supply of liquid or gaseous oxidizing agent, such as oxygen, removes electrons from the cathode as a reducing agent, such as hydrogen, supplies electrons to the anode.

Faraday's Law

Faraday's law of electrolysis can be stated as follows. The amount of a substance consumed or produced at one of the electrodes in an electrolytic cell is directly proportional to the amount of electricity that passes through the cell.

In order to use Faraday's law we need to recognize the relationship between current, time, and the amount of electric charge that flows through a circuit. By definition, one coulomb of charge is transferred when a 1-amp current flows for 1 second.

1 C = 1 amp-s

Example: To illustrate how Faraday's law can be used, let's calculate the number of grams of sodium metal that will form at the cathode when a 10.0-amp current is passed through molten sodium chloride for a period of 4.00 hours.

We start by calculating the amount of electric charge that flows through the cell.

$$10.0\ amp\ x\ 4.00\ h\ x\ \frac{60\ \text{min}}{1\ hr}\ x\ \frac{60\ s}{1\ \text{min}}\ x\ \frac{1\ c}{1\ amp-s} = 144{,}000\ c$$

Before we can use this information, we need a bridge between this macroscopic quantity and the phenomenon that occurs on the atomic scale. This bridge is represented by Faraday's constant, which describes the number of coulombs of charge carried by a mole of electrons.

$$\frac{6.022045\ x\ 10^{23}\ e^{-}}{1\ mol}\ x\ \frac{1.6021892\ x\ 10^{-19}\ c}{1\ e^{-}} = 96{,}484.56\ Clmol$$

Thus, the number of moles of electrons transferred when 144,000 coulombs of electric charge flow through the cell can be calculated as follows.

$$144{,}000\ c\ x\ \frac{1\ mol\ e^{-}}{96{,}485\ c} = 1.49\ mol\ e^{-}$$

According to the balanced equation for the reaction that occurs at the cathode of this cell, we get one mole of sodium for every mole of electrons.

Cathode (-): $Na^{+} + e^{-} \rightarrow Na$

Thus, we get 1.49 moles, or 34.3 grams, of sodium in 4.00 hours.

$$1.49\ mol\ Na\ x\ \frac{22.99\ g\ Na}{1\ mol\ Na} = 34.3\ g\ Na$$

The consequences of this calculation are interesting. We would have to run this electrolysis for more than two days to prepare a pound of sodium.

Working of an Electrolytic Cell

All electrolytic cells contain three main parts: two solid electrodes (known as the cathode and the anode) and a liquid electrolyte solution. The electrolyte solution conducts electricity because it contains dissolved ions that are free to move around throughout the solution. The cathode and anode in an electrolytic cell are attached to a source of electrical energy, like a battery.

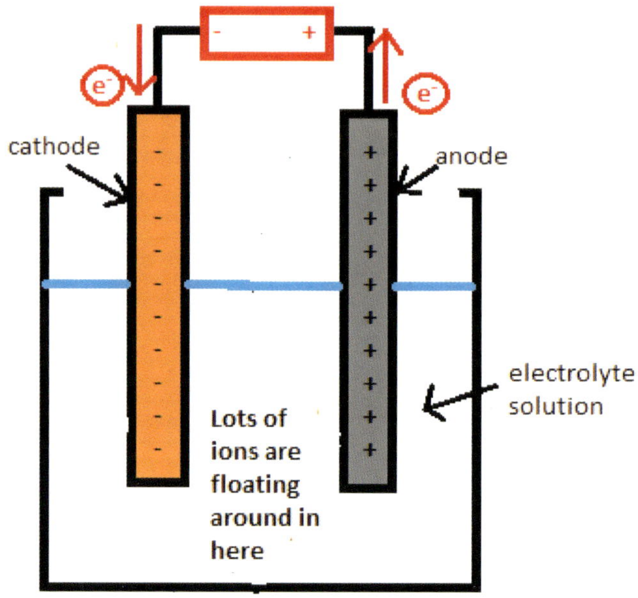

In an electrolytic cell, the cathode is always negatively charged and the anode is positively charged. These two electrodes are made out of materials that participate in the chemical reaction (such as copper, silver, and zinc). These are called active electrodes. They can also be made of chemically inert materials (like graphite, silicon, or platinum).

So, now that we know the parts, let's think about how they all work together. What actually goes on in an electrolytic cell? First, the battery provides a source of electrical energy, pushing electrons onto the cathode and making it negatively charged. Electrons are also drawn out of the anode, making it positively charged. Once this happens, an oxidation-reduction reaction is activated. At the anode, an oxidation reaction occurs, releasing electrons that then are attracted to the positively charged anode. Meanwhile, at the cathode, a reduction reaction occurs, which uses up the electrons that are building up on the cathode.

Oxidation half of the reaction: X^- --> X + e^-

Reduction half of the reaction: Y^+ + e^- --> Y

After both of these reactions, the end result is that two ions (represented here by X^- and Y^-) turn into neutral atoms or molecules (X or Y). This is possible because of the electrical energy provided by the battery that moves electrons from the anode to the cathode.

Uses

As already noted, water, particularly when ions are added (salt water or acidic water), can be *electrolyzed* (subject to electrolysis). When driven by an external source of voltage, H^+ions flow to the cathode to combine with electrons to produce hydrogen gas in a reduction reaction. Likewise, OH^-ions flow to the anode to release electrons and an H^+ion to produce oxygen gas in an oxidation reaction.

In molten sodium chloride, when a current is passed through the salt the anode oxidizes chloride ions (Cl^-) to chlorine gas, releasing electrons to the anode. Likewise the cathode reduces sodium ions (Na^+), which accept electrons from the cathode and deposits on the cathode as sodium metal.

NaCl dissolved in water can also be electrolyzed. The anode oxidizes chloride ions (Cl^-), and Cl_2 gas is produced. However, at the cathode, instead of sodium ions being reduced to sodium metal, water molecules are reduced to hydroxide ions (OH^-) and hydrogen gas (H_2). The overall result of the electrolysis is the production of chlorine gas and aqueous sodium hydroxide (NaOH) solution.

Commercially, electrolytic cells are used in electrorefining and electrowinning of several non-ferrous metals. Almost all high-purity aluminium, copper, zinc and lead is produced industrially in electrolytic cells.

Placing a piece of reactant in an electrolyte solution makes a half cell. Unless it is connected to another half cell via an electric conductor and salt bridge, no reaction will take place in a half cell.

On the cathode, reduction takes place.

Oxidant + n e^- → Reductant

Example: Cu^{2+} + 2 e → Cu

Cu^{2+} is the oxidizing agent and Cu the reducing agent.

On the anode, oxidation takes place.

Reductant → n e^- + Oxidant

Example: Zn → Zn^{2+} + 2 e^-.

Zn is the reducing agent, and Zn^{2+} the oxidizing agent.

A battery requires at least two electrodes, the anode at which oxidation occurs, and the cathode at which reduction occurs. Reduction and oxidation are always required in any battery setup.

A battery operation requires an anode, a cathode, a load, and a salt bridge (if the salt bridge is not there already). These are the key elements of a battery.

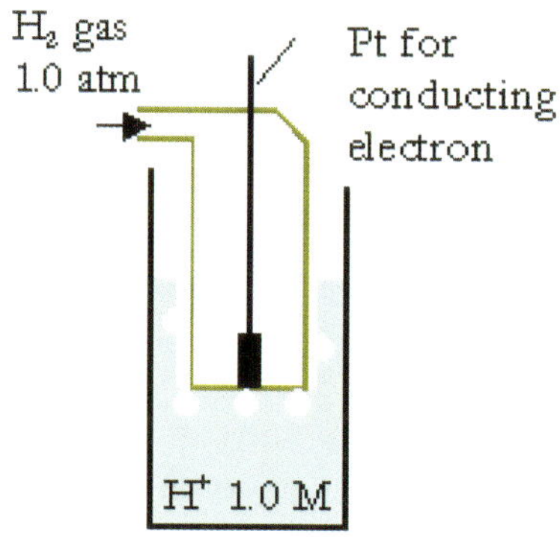

Standard Hydrogen Electrode (SHE)

The Hydrogen Half Cell

A half cell consists of an electrode and the species to be oxidized or reduced. If the material conducts electricity, it may be used as an electrode. The hydrogen electrode consists of a Pt electrode, H_2 gas and H^+. This half cell, is represented by:

$$Pt(s) \mid H_2(g) \mid H^+(aq)$$

where the vertical bars represent the phase boundaries. Conventionally, the cell potential for the hydrogen electrode is defined to be exactly zero if it has the condition as given below:

$$Pt \mid H_2\ (g, 1\ atm) \mid H^+(aq), 1\ M$$

The notations for half cells are not rigid, but a simplified way to represent a rather complicated setup.

Corrosion

Corrosion is the deterioration and loss of a material and its critical properties due to chemical, electrochemical and other reactions of the exposed material surface with the surrounding environment.

Corrosion is a dangerous and extremely costly problem. Because of it, buildings and bridges can collapse, oil pipelines break, chemical plants leak, and bathrooms flood. Corroded electrical contacts can cause fires and other problems, corroded medical implants may lead to blood poisoning, and air pollution has caused corrosion damage to works of art around the world. Corrosion threatens the safe disposal of radioactive waste that must be stored in containers for tens of thousands of years.

The most common kinds of corrosion result from electrochemical reactions. General corrosion occurs when most or all of the atoms on the same metal surface are oxidized, damaging the entire surface. Most metals are easily oxidized: they tend to lose electrons to oxygen (and other substances) in the air or in water. As oxygen is reduced (gains electrons), it forms an oxide with the metal.

When reduction and oxidation take place on different kinds of metal in contact with one another, the process is called galvanic corrosion. In electrolytic corrosion, which occurs most commonly in electronic equipment, water or other moisture becomes trapped between two electrical contacts that have an electrical voltage applied between them. The result is an unintended electrolytic cell.

Take a metal structure such as the Statue of Liberty. It looks strong and permanent. Like nearly all metal objects, however, it can become unstable as it reacts with substances in its environment and deteriorates. Sometimes this corrosion is harmless or even beneficial: the greenish patina that covers the statue's copper skin protected the metal beneath from weather damage. Inside the statue, however, corrosion caused serious harm over the years. Its iron frame and copper skin acted like the electrodes of a huge galvanic cell, so that nearly half of the frame had rusted away by 1986, the statue's one hundredth anniversary.

Galvanic Corrosion

Galvanic corrosion of aluminium. A 5-mm-thick aluminium alloy plate is physically (and hence, electrically) connected to a 10-mm-thick mild steel structural support. Galvanic corrosion occurred on the aluminium plate along the joint with the steel. Perforation of aluminium plate occurred within 2 years

Galvanic corrosion occurs when two different metals have physical or electrical contact with each other and are immersed in a common electrolyte, or when the same metal is exposed to electrolyte with different concentrations. In a galvanic couple, the more active metal (the anode) corrodes at an accelerated rate and the more noble metal (the cathode) corrodes at a slower rate. When immersed separately, each metal corrodes at its own rate. What type of metal(s) to use is readily determined by following the galvanic series. For example, zinc is often used as a sacrificial anode for steel structures. Galvanic corrosion is of major interest to the marine industry and also anywhere water (containing salts) contacts pipes or metal structures.

Factors such as relative size of anode, types of metal, and operating conditions (temperature, humidity, salinity, etc.) affect galvanic corrosion. The surface area ratio of the anode and cathode directly affects the corrosion rates of the materials. Galvanic corrosion is often prevented by the use of sacrificial anodes.

Galvanic Series

In any given environment (one standard medium is aerated, room-temperature seawater), one metal will be either more *noble* or more *active* than others, based on how strongly its ions are bound to the surface. Two metals in electrical contact share the same electrons, so that the "tug-of-war" at each surface is analogous to competition for free electrons between the two materials. Using the electrolyte as a host for the flow of ions in the same direction, the noble metal will take electrons from the active one. The resulting in mass flow or electric current can be measured to establish a hierarchy of materials in the medium of interest. This hierarchy is called a *galvanic series* and is useful in predicting and understanding corrosion.

Corrosion Removal

Often it is possible to chemically remove the products of corrosion. For example, phosphoric acid in the form of naval jelly is often applied to ferrous tools or surfaces to remove rust. Corrosion removal should not be confused with electropolishing, which removes some layers of the underlying metal to make a smooth surface. For example, phosphoric acid may also be used to electropolish copper but it does this by removing copper, not the products of copper corrosion.

Resistance to Corrosion

Some metals are more intrinsically resistant to corrosion than others. There are various ways of protecting metals from corrosion (oxidation) including painting, hot dip galvanizing, and combinations of these.

Intrinsic Chemistry

The materials most resistant to corrosion are those for which corrosion is thermodynamically unfavorable. Any corrosion products of gold or platinum tend to decompose spontaneously into pure metal, which is why these elements can be found in metallic form on Earth and have long been valued. More common "base" metals can only be protected by more temporary means.

Gold nuggets do not naturally corrode, even on a geological time scale

Some metals have naturally slow reaction kinetics, even though their corrosion is thermodynamically favorable. These include such metals as zinc, magnesium, and cadmium. While corrosion of these metals is continuous and ongoing, it happens at an acceptably slow rate. An extreme example is graphite, which releases large amounts of energy upon oxidation, but has such slow kinetics that it is effectively immune to electrochemical corrosion under normal conditions.

Passivation

Passivation refers to the spontaneous formation of an ultrathin film of corrosion products, known as a passive film, on the metal's surface that act as a barrier to further oxidation. The chemical composition and microstructure of a passive film are different from the underlying metal. Typical passive film thickness on aluminium, stainless steels, and alloys is within 10 nanometers. The passive film is different from oxide layers that are formed upon heating and are in the micrometer thickness range – the passive film recovers if removed or damaged whereas the oxide layer does not. Passivation in natural environments such as air, water and soil at moderate pH is seen in such materials as aluminium, stainless steel, titanium, and silicon.

Passivation is primarily determined by metallurgical and environmental factors. The effect of pH is summarized using Pourbaix diagrams, but many other factors are influential. Some conditions that inhibit passivation include high pH for aluminium and zinc, low pH or the presence of chloride ions for stainless steel, high temperature for titanium (in which case the oxide dissolves into the metal, rather than the electrolyte) and fluoride ions for silicon. On the other hand, unusual conditions may result in passivation of materials that are normally unprotected, as the alkaline environment of concrete does for steel rebar. Exposure to a liquid metal such as mercury or hot solder can often circumvent passivation mechanisms.

Corrosion in Passivated Materials

Passivation is extremely useful in mitigating corrosion damage, however even a high-quality alloy will corrode if its ability to form a passivating film is hindered. Proper selection of the right grade of material for the specific environment is important for the long-lasting performance of this group of materials. If breakdown occurs in the passive film due to chemical or mechanical factors, the resulting major modes of corrosion may include pitting corrosion, crevice corrosion, and stress corrosion cracking.

Pitting Corrosion

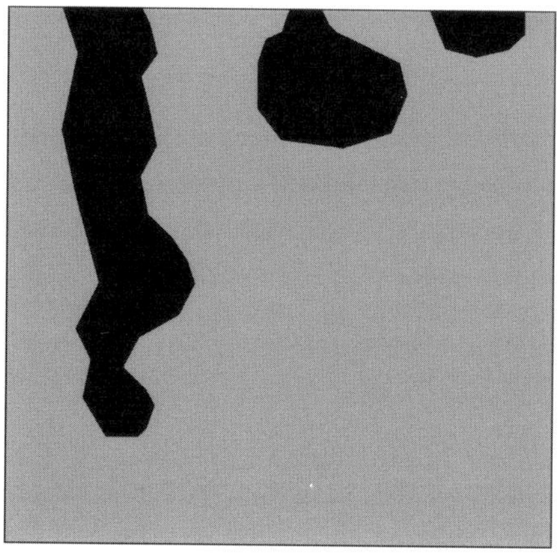

Diagram showing cross-section of pitting corrosion

Certain conditions, such as low concentrations of oxygen or high concentrations of species such as chloride which complete as anions, can interfere with a given alloy's ability to re-form a passivating film. In the worst case, almost all of the surface will remain protected, but tiny local fluctuations will degrade the oxide film in a few critical points. Corrosion at these points will be greatly amplified, and can cause *corrosion pits* of several types, depending upon conditions. While the corrosion pits only nucleate under fairly extreme circumstances, they can continue to grow even when conditions return to normal, since the interior of a pit is naturally deprived of oxygen and locally the pH decreases to very low values and the corrosion rate increases due to an autocatalytic process. In extreme cases, the sharp tips of extremely long and narrow corrosion pits can cause stress concentration to the point that otherwise tough alloys can shatter; a thin film pierced by an invisibly small hole can hide a thumb sized pit from view. These problems are especially dangerous because they are difficult to detect before a part or structure fails. Pitting remains among the most common and damaging forms of corrosion in passivated alloys, but it can be prevented by control of the alloy's environment.

Pitting results when a small hole, or cavity, forms in the metal, usually as a result of de-passivation of a small area. This area becomes anodic, while part of the remaining metal becomes cathodic, producing a localized galvanic reaction. The deterioration of this small area penetrates the metal and can lead to failure. This form of corrosion is often difficult to detect due to the fact that it is usually relatively small and may be covered and hidden by corrosion-produced compounds.

Weld Decay and Knifeline Attack

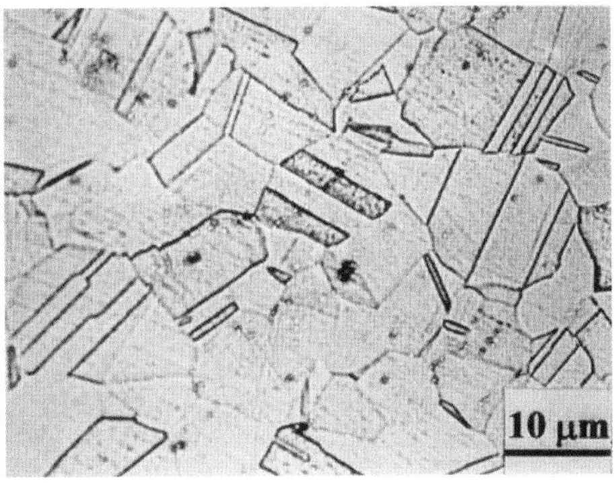

Normal microstructure of Type 304 stainless steel surface

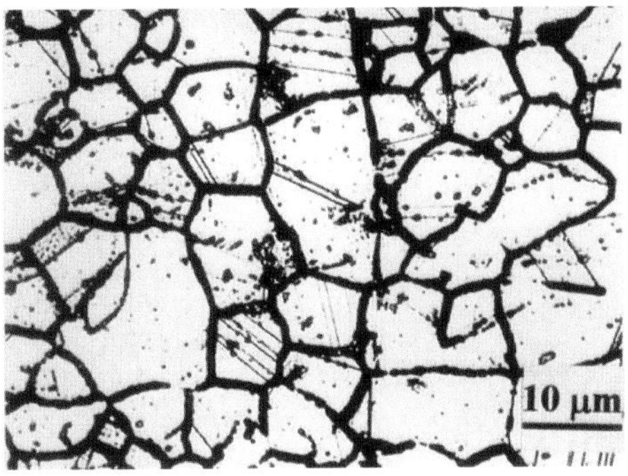

Sensitized metallic microstructure, showing wider intergranular boundaries

Stainless steel can pose special corrosion challenges, since its passivating behavior relies on the presence of a major alloying component (chromium, at least 11.5%). Because of the elevated temperatures of welding and heat treatment, chromium carbides can form in the grain boundaries of stainless alloys. This chemical reaction robs the material of chromium in the zone near the grain boundary, making those areas much less resistant to corrosion. This creates a galvanic couple with the well-protected alloy nearby, which leads to "weld decay" (corrosion of the grain boundaries in the heat affected zones) in highly corrosive environments. This process can seriously reduce the mechanical strength of welded joints over time.

A stainless steel is said to be "sensitized" if chromium carbides are formed in the microstructure. A typical microstructure of a normalized type 304 stainless steel shows no signs of sensitization, while a heavily sensitized steel shows the presence of grain boundary precipitates. The dark lines in the sensitized microstructure are networks of chromium carbides formed along the grain boundaries.

Special alloys, either with low carbon content or with added carbon "getters" such as titanium and niobium (in types 321 and 347, respectively), can prevent this effect, but the latter require special heat treatment after welding to prevent the similar phenomenon of "knifeline attack". As its name implies, corrosion is limited to a very narrow zone adjacent to the weld, often only a few micrometers across, making it even less noticeable.

Crevice Corrosion

Crevice corrosion is a localized form of corrosion occurring in confined spaces (crevices), to which the access of the working fluid from the environment is limited. Formation of a differential aeration cell leads to corrosion inside the crevices. Examples of crevices are gaps and contact areas between parts, under gaskets or seals, inside cracks and seams, spaces filled with deposits and under sludge piles.

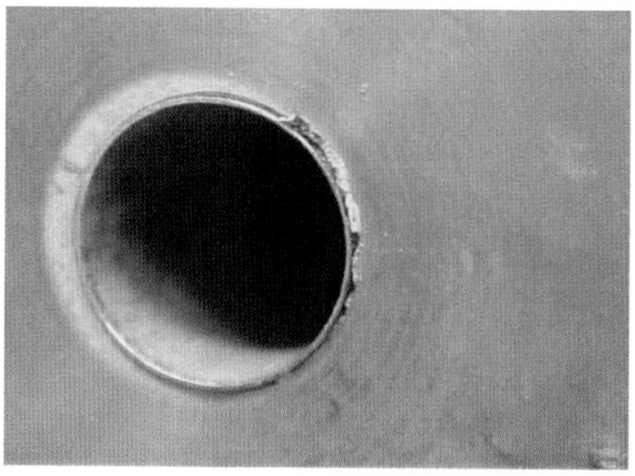

Corrosion in the crevice between the tube and tube sheet (both made of type 316 stainless steel) of a heat exchanger in a seawater desalination plant

Crevice corrosion is influenced by the crevice type (metal-metal, metal-nonmetal), crevice geometry (size, surface finish), and metallurgical and environmental factors. The susceptibility to crevice corrosion can be evaluated with ASTM standard procedures. A critical crevice corrosion temperature is commonly used to rank a material's resistance to crevice corrosion.

Microbial Corrosion

Microbial corrosion, or commonly known as microbiologically influenced corrosion (MIC), is a corrosion caused or promoted by microorganisms, usually chemoautotrophs. It can apply to both metallic and non-metallic materials, in the presence or absence of oxygen. Sulfate-reducing bacteria are active in the absence of oxygen (anaerobic); they produce hydrogen sulfide, causing sulfide stress cracking. In the presence of oxygen (aerobic), some bacteria may directly oxidize iron to iron oxides and hydroxides, other bacteria oxidize sulfur and produce sulfuric acid causing biogenic sulfide corrosion. Concentration cells can form in the deposits of corrosion products, leading to localized corrosion.

Accelerated low-water corrosion (ALWC) is a particularly aggressive form of MIC that affects steel piles in seawater near the low water tide mark. It is characterized by an orange sludge, which smells of hydrogen sulfide when treated with acid. Corrosion rates can be very high and design corrosion allowances can soon be exceeded leading to premature failure of the steel pile. Piles that have been coated and have cathodic protection installed at the time of construction are not susceptible to ALWC. For unprotected piles, sacrificial anodes can be installed locally to the affected areas to inhibit the corrosion or a complete retrofitted sacrificial anode system can be installed. Affected areas can also be treated using cathodic protection, using either sacrificial anodes or applying current to an inert anode to produce a calcareous deposit, which will help shield the metal from further attack.

High-temperature Corrosion

High-temperature corrosion is chemical deterioration of a material (typically a metal) as a result of heating. This non-galvanic form of corrosion can occur when a metal is subjected to a hot atmosphere containing oxygen, sulfur, or other compounds capable of oxidizing (or assisting the oxidation of) the material concerned. For example, materials used in aerospace, power generation and even in car engines have to resist sustained periods at high temperature in which they may be exposed to an atmosphere containing potentially highly corrosive products of combustion.

The products of high-temperature corrosion can potentially be turned to the advantage of the engineer. The formation of oxides on stainless steels, for example, can provide a protective layer preventing further atmospheric attack, allowing for a material to be used for sustained periods at both room and high temperatures in hostile conditions. Such high-temperature corrosion products, in the form of compacted oxide layer glazes, prevent or reduce wear during high-temperature sliding contact of metallic (or metallic and ceramic) surfaces.

Metal Dusting

Metal dusting is a catastrophic form of corrosion that occurs when susceptible materials are exposed to environments with high carbon activities, such as synthesis gas and other high-CO environments. The corrosion manifests itself as a break-up of bulk metal to metal powder. The suspected mechanism is firstly the deposition of a graphite layer on the surface of the metal, usually from carbon monoxide (CO) in the vapor phase. This graphite layer is then thought to form metastable M_3C species (where M is the metal), which migrate away from the metal surface. However, in some regimes no M_3C species is observed indicating a direct transfer of metal atoms into the graphite layer.

Protection from Corrosion

The US military shrink wraps equipment such as helicopters to protect them from corrosion and thus save millions of dollars

Various treatments are used to slow corrosion damage to metallic objects which are exposed to the weather, salt water, acids, or other hostile environments. Some unprotected metallic alloys are extremely vulnerable to corrosion, such as those used in neodymium magnets, which can spall or crumble into powder even in dry, temperature-stable indoor environments unless properly treated to discourage corrosion.

Surface Treatments

When surface treatments are used to retard corrosion, great care must be taken to ensure complete coverage, without gaps, cracks, or pinhole defects. Small defects can act as an "Achilles' heel", allowing corrosion to penetrate the interior and causing extensive damage even while the outer protective layer remains apparently intact for a period of time.

Applied Coatings

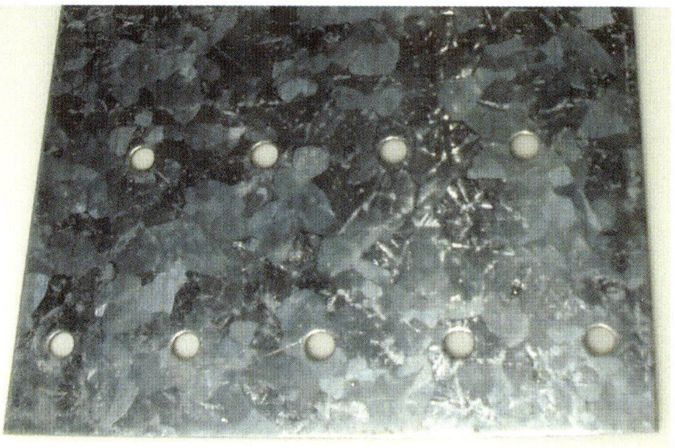

Galvanized surface

Plating, painting, and the application of enamel are the most common anti-corrosion treatments. They work by providing a barrier of corrosion-resistant material between the damaging environment and the structural material. Aside from cosmetic and manufacturing issues, there may be tradeoffs in mechanical flexibility versus resistance to abrasion and high temperature. Platings usually fail only in small sections, but if the plating is more noble than the substrate (for example, chromium on steel), a galvanic couple will cause any exposed area to corrode much more rapidly than an unplated surface would. For this reason, it is often wise to plate with active metal such as zinc or cadmium.

Painting either by roller or brush is more desirable for tight spaces; spray would be better for larger coating areas such as steel decks and waterfront applications. Flexible polyurethane coatings, like Durabak-M26 for example, can provide an anti-corrosive seal with a highly durable slip resistant membrane. Painted coatings are relatively easy to apply and have fast drying times although temperature and humidity may cause dry times to vary.

Reactive Coatings

If the environment is controlled (especially in recirculating systems), corrosion inhibitors can often be added to it. These chemicals form an electrically insulating or chemically impermeable coating on exposed metal surfaces, to suppress electrochemical reactions. Such methods make the system less sensitive to scratches or defects in the coating, since extra inhibitors can be made available wherever metal becomes exposed.

Chemicals that inhibit corrosion include some of the salts in hard water (Roman water systems are famous for their mineral deposits), chromates, phosphates, polyaniline, other conducting polymers and a wide range of specially-designed chemicals that resemble surfactants (i.e. long-chain organic molecules with ionic end groups).

Anodization

This climbing descender is anodized with a yellow finish

Aluminium alloys often undergo a surface treatment. Electrochemical conditions in the bath are carefully adjusted so that uniform pores, several nanometers wide, appear in the metal's oxide film. These pores allow the oxide to grow much thicker than passivating conditions would allow. At the end of the treatment, the pores are allowed to seal, forming a harder-than-usual surface layer. If this coating is scratched, normal passivation processes take over to protect the damaged area.

Anodizing is very resilient to weathering and corrosion, so it is commonly used for building facades and other areas where the surface will come into regular contact with the elements. While being resilient, it must be cleaned frequently. If left without cleaning, panel edge staining will naturally occur. Anodization is the process of converting an anode into cathode by bringing a more active anode in contact with it.

Biofilm Coatings

A new form of protection has been developed by applying certain species of bacterial films to the surface of metals in highly corrosive environments. This process increases the corrosion resistance substantially. Alternatively, antimicrobial-producing biofilms can be used to inhibit mild steel corrosion from sulfate-reducing bacteria.

Controlled Permeability Formwork

Controlled permeability formwork (CPF) is a method of preventing the corrosion of reinforcement by naturally enhancing the durability of the cover during concrete placement. CPF has been used in environments to combat the effects of carbonation, chlorides, frost and abrasion.

Cathodic Protection

Cathodic protection (CP) is a technique to control the corrosion of a metal surface by making that surface the cathode of an electrochemical cell. Cathodic protection systems are most commonly used to protect steel, and pipelines and tanks; steel pier piles, ships, and offshore oil platforms.

Sacrificial Anode Protection

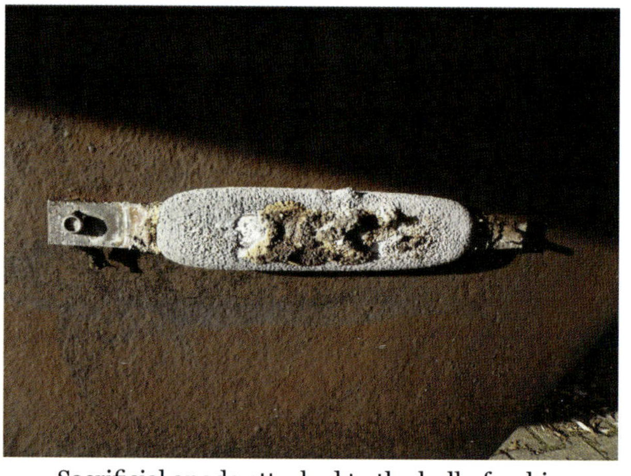

Sacrificial anode attached to the hull of a ship

For effective CP, the potential of the steel surface is polarized (pushed) more negative until the metal surface has a uniform potential. With a uniform potential, the driving force for the corrosion reaction is halted. For galvanic CP systems, the anode material corrodes under the influence of the steel, and eventually it must be replaced. The polarization is caused by the current flow from the anode to the cathode, driven by the difference in electrode potential between the anode and the cathode.

Impressed Current Cathodic Protection

For larger structures, galvanic anodes cannot economically deliver enough current to provide complete protection. Impressed current cathodic protection (ICCP) systems use anodes connected to a DC power source (such as a cathodic protection rectifier). Anodes for ICCP systems are tubular and solid rod shapes of various specialized materials. These include high silicon cast iron, graphite, mixed metal oxide or platinum coated titanium or niobium coated rod and wires.

Anodic Protection

Anodic protection impresses anodic current on the structure to be protected (opposite to the cathodic protection). It is appropriate for metals that exhibit passivity (e.g. stainless steel) and suitably small passive current over a wide range of potentials. It is used in aggressive environments, such as solutions of sulfuric acid.

Rate of Corrosion

These neodymium magnets corroded extremely rapidly after only 5 months of outside exposure

A simple test for measuring corrosion is the weight loss method. The method involves exposing a clean weighed piece of the metal or alloy to the corrosive environment for a specified time followed by cleaning to remove corrosion products and weighing the piece to determine the loss of weight. The rate of corrosion (R) is calculated as,

$$R = \frac{kW}{\rho At}$$

where k is a constant, W is the weight loss of the metal in time t, A is the surface area of the metal exposed, and ρ is the density of the metal (in g/cm^3).

Other common expressions for the corrosion rate is penetration depth and change of mechanical properties.

Economic Impact

The collapsed Silver Bridge, as seen from the Ohio side

In 2002, the US Federal Highway Administration released a study titled "Corrosion Costs and Preventive Strategies in the United States" on the direct costs associated with metallic corrosion in the US industry. In 1998, the total annual direct cost of corrosion in the U.S. was ca. \$276 billion (ca. 3.2% of the US gross domestic product). Broken down into five specific industries, the economic losses are \$22.6 billion in infrastructure; $17.6 billion in production and manufacturing; $29.7 billion in transportation; $20.1 billion in government; and $47.9 billion in utilities.

Rust is one of the most common causes of bridge accidents. As rust has a much higher volume than the originating mass of iron, its build-up can also cause failure by forcing apart adjacent parts. It was the cause of the collapse of the Mianus river bridge in 1983, when the bearings rusted internally and pushed one corner of the road slab off its support. Three drivers on the roadway at the time died as the slab fell into the river below. The following NTSB investigation showed that a drain in the road had been blocked for road re-surfacing, and had not been unblocked; as a result, runoff water penetrated the support hangers. Rust was also an important factor in the Silver Bridge disaster of 1967 in West Virginia, when a steel suspension bridge collapsed within a minute, killing 46 drivers and passengers on the bridge at the time.

Similarly, corrosion of concrete-covered steel and iron can cause the concrete to spall, creating severe structural problems. It is one of the most common failure modes of reinforced concrete bridges. Measuring instruments based on the half-cell potential can detect the potential corrosion spots before total failure of the concrete structure is reached.

Until 20–30 years ago, galvanized steel pipe was used extensively in the potable water systems for single and multi-family residents as well as commercial and public construction. Today, these systems have long ago consumed the protective zinc and are corroding internally resulting in poor water quality and pipe failures. The economic impact on homeowners, condo dwellers, and the public infrastructure is estimated at 22 billion dollars as the insurance industry braces for a wave of claims due to pipe failures.

Corrosion in Nonmetals

Most ceramic materials are almost entirely immune to corrosion. The strong chemical bonds that hold them together leave very little free chemical energy in the structure; they can be thought of as already corroded. When corrosion does occur, it is almost always a simple dissolution of the material or chemical reaction, rather than an electrochemical process. A common example of corrosion protection in ceramics is the lime added to soda-lime glass to reduce its solubility in water; though it is not nearly as soluble as pure sodium silicate, normal glass does form sub-microscopic flaws when exposed to moisture. Due to its brittleness, such flaws cause a dramatic reduction in the strength of a glass object during its first few hours at room temperature.

Corrosion of Polymers

Polymer degradation involves several complex and often poorly understood physiochemical processes. These are strikingly different from the other processes discussed here, and so the term "corrosion" is only applied to them in a loose sense of the word. Because of their large molecular weight, very little entropy can be gained by mixing a given mass of polymer with another substance, making them generally quite difficult to dissolve. While dissolution is a problem in some polymer applications, it is relatively simple to design against.

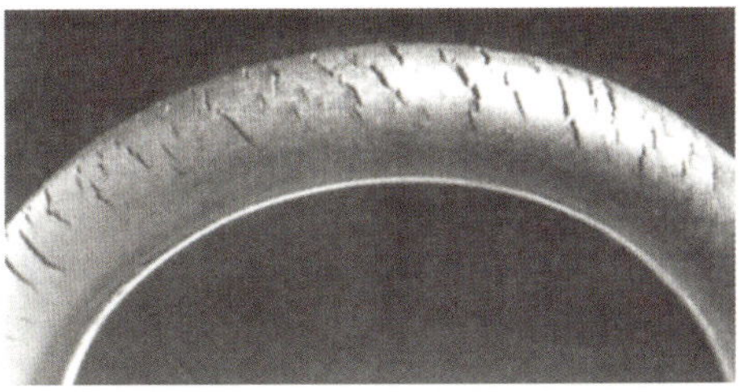

Ozone cracking in natural rubber tubing

A more common and related problem is "swelling", where small molecules infiltrate the structure, reducing strength and stiffness and causing a volume change. Conversely, many polymers (notably flexible vinyl) are intentionally swelled with plasticizers, which can be leached out of the structure, causing brittleness or other undesirable changes.

The most common form of degradation, however, is a decrease in polymer chain length. Mechanisms which break polymer chains are familiar to biologists because of their effect on DNA: ionizing radiation (most commonly ultraviolet light), free radicals, and oxidizers such as oxygen, ozone, and chlorine. Ozone cracking is a well-known problem affecting natural rubber for example. Plastic additives can slow these process very effectively, and can be as simple as a UV-absorbing pigment (e.g. titanium dioxide or carbon black). Plastic shopping bags often do not include these additives so that they break down more easily as ultrafine particles of litter.

Corrosion of Glasses

Glass corrosion

Glass is characterized by a high degree of corrosion-resistance. Because of its high water-resistance it is often used as primary packaging material in the pharma industry since most medicines are preserved in a watery solution. Besides its water-resistance, glass is also very robust when being exposed to chemically aggressive liquids or gases. While other materials like metal or plastics quickly reach their limits, special glass-types can easily hold up.

Glass disease is the corrosion of silicate glasses in aqueous solutions. It is governed by two mechanisms: diffusion-controlled leaching (ion exchange) and hydrolytic dissolution of the glass network. Both mechanisms strongly depend on the pH of contacting solution: the rate of ion exchange decreases with pH as $10^{-0.5pH}$ whereas the rate of hydrolytic dissolution increases with pH as $10^{0.5pH}$.

Mathematically, corrosion rates of glasses are characterized by normalized corrosion rates of elements NR_i (g/cm²·d) which are determined as the ratio of total amount of released species into the water M_i (g) to the water-contacting surface area S (cm²), time of contact t (days) and weight fraction content of the element in the glass f_i:

$$\mathrm{NR}_i = \frac{M_i}{Sf_i t}.$$

The overall corrosion rate is a sum of contributions from both mechanisms (leaching + dissolution) $NR_i = NRx_i + NRh$. Diffusion-controlled leaching (ion exchange) is characteristic of the initial phase of corrosion and involves replacement of alkali ions in the glass by a hydronium (H_3O^+) ion from the solution. It causes an ion-selective depletion of near surface layers of glasses and gives an inverse square root dependence of corrosion rate with exposure time. The diffusion-controlled normalized leaching rate of cations from glasses (g/cm²·d) is given by:

$$\mathrm{NR}x_i = 2\rho\sqrt{\frac{D_i}{\pi t}},$$

where t is time, D_i is the i-th cation effective diffusion coefficient (cm²/d), which depends on pH of contacting water as $D_i = D_{io} \cdot 10^{-pH}$, and ρ is the density of the glass (g/cm³).

Glass network dissolution is characteristic of the later phases of corrosion and causes a congruent release of ions into the water solution at a time-independent rate in dilute solutions (g/cm²·d):

$$\mathrm{NR}h = \rho r_h,$$

where r_h is the stationary hydrolysis (dissolution) rate of the glass (cm/d). In closed systems the consumption of protons from the aqueous phase increases the pH and causes a fast transition to hydrolysis. However, a further saturation of solution with silica impedes hydrolysis and causes the glass to return to an ion-exchange, e.g. diffusion-controlled regime of corrosion.

In typical natural conditions normalized corrosion rates of silicate glasses are very low and are of the order of 10^{-7}–10^{-5} g/(cm²·d). The very high durability of silicate glasses in water makes them suitable for hazardous and nuclear waste immobilisation.

Glass Corrosion Tests

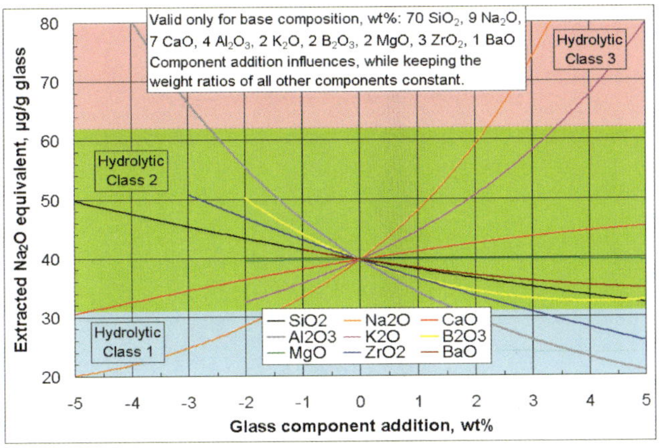

Effect of addition of a certain glass component on the chemical durability against water corrosion of a specific base glass (corrosion test ISO 719)

There exist numerous standardized procedures for measuring the corrosion (also called chemical durability) of glasses in neutral, basic, and acidic environments, under simulated environmental conditions, in simulated body fluid, at high temperature and pressure, and under other conditions.

The standard procedure ISO 719 describes a test of the extraction of water-soluble basic compounds under neutral conditions: 2 g of glass, particle size 300–500 µm, is kept for 60 min in 50 ml de-ionized water of grade 2 at 98°C; 25 ml of the obtained solution is titrated against 0.01 mol/l HCl solution. The volume of HCl required for neutralization is classified according to the table below.

Amount of 0.01M HCl needed to neutralize extracted basic oxides, ml	**Extracted Na_2O equivalent, µg**	**Hydrolytic class**
< 0.1	< 31	1
0.1-0.2	31-62	2
0.2-0.85	62-264	3
0.85-2.0	264-620	4
2.0-3.5	620-1085	5
> 3.5	> 1085	> 5

The standardized test ISO 719 is not suitable for glasses with poor or not extractable alkaline components, but which are still attacked by water, e.g. quartz glass, B_2O_3 glass or P_2O_5 glass.

Usual glasses are differentiated into the following classes:

- Hydrolytic class 1 (Type I): This class, which is also called neutral glass, includes borosilicate glasses (e.g. Duran, Pyrex, Fiolax).

  Glass of this class contains essential quantities of boron oxides, aluminium oxides and alkaline earth oxides. Through its composition neutral glass has a high resistance against temperature shocks and the highest hydrolytic resistance. Against acid and neutral solutions it shows high chemical resistance, because of its poor alkali content against alkaline solutions.

- Hydrolytic class 2 (Type II): This class usually contains sodium silicate glasses with a high hydrolytic resistance through surface finishing. Sodium silicate glass is a silicate glass, which contains alkali- and alkaline earth oxide and primarily sodium oxide and Calcium oxide.

- Hydrolytic class 3 (Type III): Glass of the 3rd hydrolytic class usually contains sodium silicate glasses and has a mean hydrolytic resistance, which is two times poorer than of type 1 glasses.

  Acid class DIN 12116 and alkali class DIN 52322 (ISO 695) are to be distinguished from the hydrolytic class DIN 12111 (ISO 719).

Electrolysis

Electrolysis, process by which electric current is passed through a substance to effect a chemical change. The chemical change is one in which the substance loses or gains an electron (oxidation or reduction). The process is carried out in an electrolytic cell (q.v.), an apparatus consisting of positive and negative electrodes held apart and dipped into a solution containing positively and negatively charged ions. The substance to be transformed may form the electrode, may constitute the solution, or may be dissolved in the solution. Electric current (i.e., electrons) enters through the negatively charged electrode (cathode); positively charged components of the solution travel to this electrode, combine with the electrons, and are transformed to neutral elements or molecules. The negatively charged components of the solution travel to the other electrode (anode), give up their electrons, and are transformed into neutral elements or molecules. If the substance to be transformed is the electrode, the reaction is generally one in which the electrode dissolves by giving up electrons.

Electrolysis is used extensively in metallurgical processes, such as in extraction (electrowinning) or purification (electrorefining) of metals from ores or compounds and in deposition of metals from solution (electroplating). Metallic sodium and chlorine gas are produced by the electrolysis of molten sodium chloride; electrolysis of an aqueous solution of sodium chloride yields sodium hydroxide and chlorine gas. Hydrogen and oxygen are produced by the electrolysis of water.

The main components required to achieve electrolysis are:

- An electrolyte: a substance containing free ions, which are the carriers of electric current in the electrolyte. If the ions are not mobile, as in a solid salt, then electrolysis cannot occur.
- A direct current (DC) supply: provides the energy necessary to create or discharge the ions in the electrolyte. Electric current is carried by electrons in the external circuit.
- Two electrodes: an electrical conductor that provides the physical interface between the electrical circuit providing the energy and the electrolyte.

The Interchange of Atoms and Ions

The key process of electrolysis is the interchange of atoms and ions by the removal or addition of electrons to the external circuit. The required products of electrolysis are in a different physical state from the electrolyte and can be removed by some physical processes.

Each electrode attracts ions that are of the opposite charge. Positively charged ions, or cations, move toward the electron-providing cathode, which is negative; negatively charged ions, or anions, move toward the positive anode. You may have noticed that this is the opposite of a galvanic cell, where the anode is negative and the cathode is positive.

At the electrodes, electrons are absorbed or released by the atoms and ions. Those atoms that gain or lose electrons become charged ions that pass into the electrolyte. Those ions that gain or lose electrons to become uncharged atoms separate from the electrolyte. The formation of uncharged atoms from ions is called discharging. The energy required to cause the ions to migrate to the electrodes, and the energy to cause the change in ionic state, is provided by the external source.

Oxidation and Reduction

Oxidation of ions or neutral molecules occurs at the anode, and reduction of ions or neutral molecules occurs at the cathode. For example, it is possible to oxidize ferrous ions to ferric ions at the anode:

$$Fe^{2+}(aq) \rightarrow Fe^{3+}(aq) + e^-$$

It is also possible to reduce ferricyanide ions to ferrocyanide ions at the cathode:

$$Fe(CN)_6^{3-} + e^- \rightarrow Fe(CN)_6^{4-}$$

Neutral molecules can also react at either electrode. Electrolysis reactions involving H^+ ions are fairly common in acidic solutions. In alkaline water solutions, reactions involving hydroxide ions (OH^-) are common. The substances oxidized or reduced can also be the solvent, which is usually water, or the electrodes. It is possible to have electrolysis involving gases.

Predicting the Products of Electrolysis

Let's look at how to predict the products. For example, what two ions will $CuSO_4$ break down into? The answer is Cu^{2+} and SO_4^{2-} . Let's look more closely at this reaction.

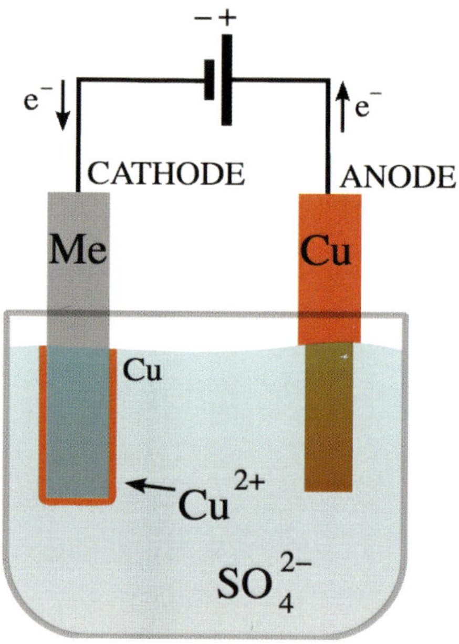

Electrolysis of copper sulfate: Two copper electrodes are placed in a solution of blue copper sulfate and are connected to a source of electrical current. The current is turned on for a period of time

We take two copper electrodes and place them into a solution of blue copper sulfate ($CuSO_4$) and then turn the current on. We notice that the the initial blue color of the solution remains unchanged, but it appears that copper has been deposited on one of the electrodes but dissolved on the other. This is because Cu^{2+} ions are attracted to the negatively charged cathode, and since the the cathode is putting out electrons, the Cu^{2+} becomes reduced to form copper metal, which is deposited on the electrode. The reaction at this electrode is:

$$Cu^{2+}(aq) + 2e-\rightarrow Cu(s)$$

At the positive anode, copper metal is oxidized to form Cu^{2+} ions. This is why it appears that the copper has dissolved from the electrode. The reaction at this electrode is:

$$Cu(s)\rightarrow Cu^{2+}(aq) + 2e^-$$

We just saw electric current used to split $CuSO_4$ into its component ions. This is all it takes to predict the products of electrolysis; all you have to do is break down a compound into its component ions.

Electrolysis of Sodium Chloride

Two commonly used methods of electrolysis involve molten sodium chloride and aqueous sodium chloride, which give different products.

Electrolysis of NaCl

As we have covered, electrolysis is the passage of a direct electric current through an ionic substance that is either molten or dissolved in a suitable solvent. This results in chemical reactions at the electrodes and the separation of materials. Two commonly used methods of electrolysis involve molten sodium chloride and aqueous sodium chloride. You might think that both methods would give you the same products, but this not the case. Let's go through each of the methods to understand the different processes.

Electrolysis of Molten NaCl

If sodium chloride is melted (above 801 °C), two electrodes are inserted into the melt, and an electric current is passed through the molten salt, then chemical reactions take place at the electrodes.

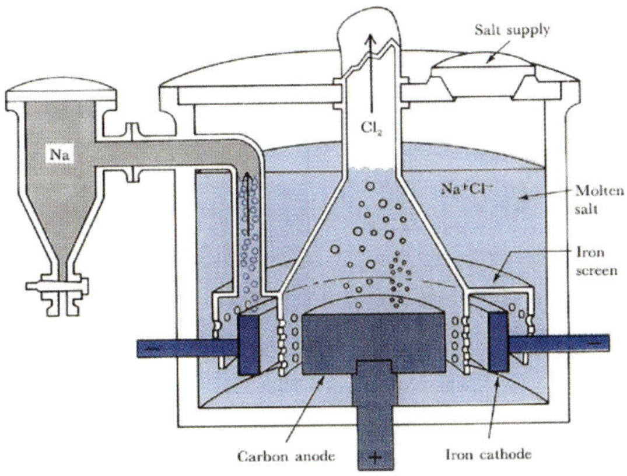

Electrolysis cell for molten sodium chloride

A commercial electrolysis cell for the production of metallic sodium and chlorine gas from molten NaCl. Liquid sodium floats to the top of the melt above the cathode and is drained off into a storage tank. Chlorine gas bubbles out of the melt above the anode.

Sodium ions migrate to the cathode, where electrons enter the melt and are reduced to sodium metal:

$$Na^+ + e- \rightarrow Na$$

Chloride ions migrate the other way, toward the anode. They give up their electrons to the anode and are oxidized to chlorine gas:

$$CI^{-} \rightarrow \frac{1}{2}CI_{2} + e^{-}$$

The overall reaction is the breakdown of sodium chloride into its elements:

$$2NaCl \rightarrow 2Na(s) + Cl_{2}\ (g)$$

Electrolysis of Aqueous NaCl

What happens when we have an aqueous solution of sodium chloride? Well, we can't forget that we have to factor water into the equation. Since water can be both oxidized and reduced, it competes with the dissolved Na^{+} and Cl^{-} ions. Rather than producing sodium, hydrogen is produced.

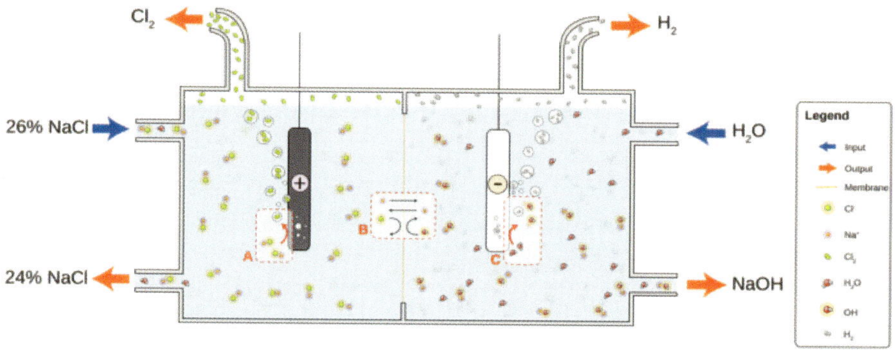

Electrolysis of aqueous sodium chloride

Electrolysis of aqueous NaCl results in hydrogen and chloride gas. At the anode (A), chloride (Cl-) is oxidized to chlorine. The ion-selective membrane (B) allows the counterion Na+ to freely flow across, but prevents anions such as hydroxide (OH-) and chloride from diffusing across. At the cathode (C), water is reduced to hydroxide and hydrogen gas. The net process is the electrolysis of an aqueous solution of NaCl into industrially useful products sodium hydroxide (NaOH) and chlorine gas.

The reaction at the cathode is:

$$H_{2}O(l) + 2e^{-} \rightarrow H_{2}(g) + 2OH^{-}$$

The reaction at the anode is:

$$CI^{-} \rightarrow \frac{1}{2}CI_{2}(g) + 1e^{-}$$

The overall reaction is as follows:

$$NaCI(aq) + H_2O(1) \rightarrow Na^+(aq) + OH^-(aq) + H_2(g) + \frac{1}{2}CI_2(g)$$

Reduction of Na^+ ($E^\circ = -2.7$ v) is energetically more difficult than the reduction of water (-1.23 v), so in aqueous solution, the latter will prevail.

Deduce the products of the electrolysis of a molten salt: Electrolysis of a molten salt produces the elements from the salt. So, the electrolysis of WCl4 produces W and Cl2. Metal ions receive electrons at the negative electrode, and the non-metals lose them at the positive electrode.

Electrolysis of Water

Pure water cannot undergo significant electrolysis without an electrolyte, such as an acid or a base.

Pure water cannot undergo significant electrolysis without adding an electrolyte. If the object is to produce hydrogen and oxygen, the added electrolyte must be energetically more difficult to oxidize or reduce than water itself. For example, electrolysis of a solution of sulfuric acid or of a salt, such as $NaNO_3$, results in the decomposition of water at both electrodes:

Cathode: $H_2O+2e^- \rightarrow H_2(g)+2OH^-$

$E = 0.00$ V

Anode: $2H_2O \rightarrow O_2(g)+4H^++4e^-$

$E^\circ = -1.23$ V

Multiplying the cathode reaction by 2, in order to match the number of electrons transferred, results in this net equation, after OH^- and H^+ ions combine to form water:

Net: 2H2O(l)→2H2(g)+O2(g)

$E = -1.23$ v

Hydrogen will appear at the cathode, the negatively charged electrode, where electrons enter the water, and oxygen will appear at the anode, the positively charged electrode. The number of moles of hydrogen generated is twice the number of moles of oxygen, and both are proportional to the total electrical charge conducted by the solution. The number of electrons pushed through the water is twice the number of generated hydrogen molecules, and four times the number of generated oxygen molecules.

Johann Ritter, who went on to invent the first electrochemical cell, was one of the first people to discover the decomposition of water by electricity.

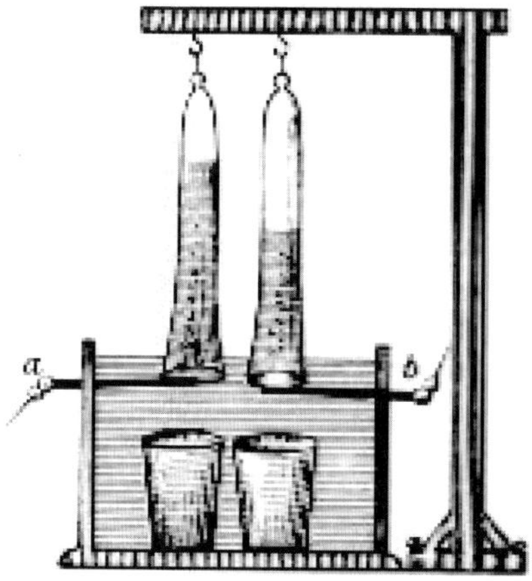

Electrolysis of water: Device invented by Johann Wilhelm Ritter to develop the electrolysis of water

Electrolysis Stoichiometry

The amount of chemical change that occurs in electrolysis is stoichiometrically related to the amount of electrons that pass through the cell.

Stoichiometry of an Electrolytic Cell

The extent of chemical change that occurs in an electrolytic cell is stoichiometrically related to the number of moles of electrons that pass through the cell. From the perspective of the voltage source and circuit outside the electrodes, the flow of electrons is generally described in terms of electrical current using the SI units of coulombs and amperes. It takes 96,485 coulombs to constitute a mole of electrons, a unit known as the faraday (F).

This relation was first formulated by Michael Faraday in 1832, in the form of two laws of electrolysis:

1. The weights of substances formed at an electrode during electrolysis are directly proportional to the quantity of electricity that passes through the electrolyte.
2. The weights of different substances formed by the passage of the same quantity of electricity are proportional to the equivalent weight of each substance.

The equivalent weight of a substance is defined as the molar mass divided by the number of electrons required to oxidize or reduce each unit of the substance. Thus, one mole of V^{3+} corresponds to three equivalents of this species, and will require three faradays of charge to deposit it as metallic vanadium. $\left(V^{3+} + 3e^{-} \rightarrow V\right)$.

Most stoichiometric problems involving electrolysis can be solved without explicit use of Faraday's laws. The "chemistry" in these problems is usually very elementary; the major difficulties usually stem from unfamiliarity with the basic electrical units:

- Current (in amperes) is the rate of charge transport: $1 \text{ amp} = 1 \frac{\text{Coulombs}}{\text{second}}$
- Power (in watts) is the rate of energy production or consumption: $1 \text{ w} = 1 \frac{\text{Joule}}{\text{second}}$
- 1 Kilowatt-hour = 3600 J.

References

- Hudlický, Miloš (1996). Reductions in Organic Chemistry. Washington, D.C.: American Chemical Society. p. 429. ISBN 0-8412-3344-6
- Electrochemical-cells, boundless-chemistry: lumenlearning.com, Retrieved 29 June 2018
- R. Zuo; D. Örnek; B.C. Syrett; R.M. Green; C.-H. Hsu; F.B. Mansfeld; T.K. Wood (2004). "Inhibiting mild steel corrosion from sulfate-reducing bacteria using antimicrobial-producing biofilms in Three-Mile-Island process water". Appl. Microbiol. Biotechnol. 64 (2): 275–283. doi:10.1007/s00253-003-1403-7
- Electrochemical-cell, physical-chemistry: chemistry.tutorvista.com, Retrieved 31 March 2018
- Daniel Robles. "Potable Water Pipe Condition Assessment For a High Rise Condominium in The Pacific Northwest". GSG Group, Inc. Dan Robles, PE. Retrieved 10 December 2012
- General-knowledge, What-is-galvanic-cells-and-how-it-work-1456202213-1: Retrieved 30 May 2018
- Hudlický, Miloš (1990). Oxidations in Organic Chemistry. Washington, D.C.: American Chemical Society. p. 456. ISBN 0-8412-1780-7
- Oxidation-reduction-reaction, science: britannica.com, Retrieved 11 July 2018
- Vapor Hydration Testing (VHT) Archived December 14, 2007, at the Wayback Machine.. Vscht.cz. Retrieved on 2012-07-15
- Corrosion-science: electrochem.org, Retrieved 28 March 2018
- Phillips, John; Strozak, Victor; Wistrom, Cheryl (2000). Chemistry: Concepts and Applications. Glencoe McGraw-Hill. p. 558. ISBN 978-0-02-828210-7

6

Chemical Kinetics

Chemical kinetics studies the rates of chemical reactions and processes. It is focused on an investigation of reaction mechanisms, transition states as well as the factors that influence reaction rates. All the diverse aspects of chemical kinetics have been carefully analyzed in this chapter, such as catalysis, reaction mechanisms and reaction rates, besides others.

Chemical Kinetics is the branch of physical chemistry that is concerned with understanding the rates of chemical reactions. It is to be contrasted with thermodynamics, which deals with the direction in which a process occurs but in itself tells nothing about its rate. Thermodynamics is time's arrow, while chemical kinetics is time's clock. Chemical kinetics relates to many aspects of cosmology, geology, biology, engineering, and even psychology and thus has far-reaching implications. The principles of chemical kinetics apply to purely physical processes as well as to chemical reactions.

One reason for the importance of kinetics is that it provides evidence for the mechanisms of chemical processes. Besides being of intrinsic scientific interest, knowledge of reaction mechanisms is of practical use in deciding what is the most effective way of causing a reaction to occur. Many commercial processes can take place by alternative reaction paths, and knowledge of the mechanisms makes it possible to choose reaction conditions that favour one path over others.

A chemical reaction is, by definition, one in which chemical substances are transformed into other substances, which means that chemical bonds are broken and formed so that there are changes in the relative positions of atoms in molecules. At the same time, there are shifts in the arrangements of the electrons that form the chemical bonds. A description of a reaction mechanism must therefore deal with the movements and speeds of atoms and electrons. The detailed mechanism by which a chemical process occurs is referred to as the reaction path, or pathway.

The vast amount of work done in chemical kinetics has led to the conclusion that some chemical reactions go in a single step; these are known as elementary reactions. Other reactions go in more than one step and are said to be stepwise, composite, or complex. Measurements of the rates of chemical reactions over a range of conditions can show whether a reaction proceeds by one or more steps. If a reaction is stepwise, kinetic measurements provide evidence for the mechanism of the individual elementary steps. Information about reaction mechanisms is also provided by certain non-kinetic studies, but little can be known about a mechanism until its kinetics has been investigated.

Even then, some doubt must always remain about a reaction mechanism. An investigation, kinetic or otherwise, can disprove a mechanism but can never establish it with absolute certainty.

Reaction Rates

Chemical reaction rates are the rates of change in concentrations or amounts of either reactants or products. For changes in amounts, the units can be one of mol/s, g/s, lb/s, kg/day etc. For changes in concentrations, the units can be one of mol/(L s), g/(L s), %/s etc.

With respect to reaction rates, we may deal with *average rates, instantaneous rates, or initial rates* depending on the experimental conditions.

Thermodynamics and kinetics are two factors that affect reaction rates. The study of energy gained or released in chemical reactions is called thermodynamics, and such energy data are called thermodynamic data. However, thermodynamic data have no direct correlation with reaction rates, for which the kinetic factor is perhaps more important. For example, at room temperature (a wide range of temperatures), thermodynamic data indicates that diamond shall convert to graphite, but in reality, the conversion rate is so slow that most people think that *diamond is forever*.

Factors Affecting Reaction Rate

Nature of the Reactants

The reaction rate varies depending upon what substances are reacting. Acid/base reactions, the formation of salts, and ion exchange are usually fast reactions. When covalent bond formation takes place between the molecules and when large molecules are formed, the reactions tend to be slower.

The nature and strength of bonds in reactant molecules greatly influence the rate of their transformation into products.

Physical State

The physical state (solid, liquid, or gas) of a reactant is also an important factor of the rate of change. When reactants are in the same phase, as in aqueous solution, thermal motion brings them into contact. However, when they are in different phases, the reaction is limited to the interface between the reactants. Reaction can occur only at their area of contact; in the case of a liquid and a gas, at the surface of the liquid. Vigorous shaking and stirring may be needed to bring the reaction to completion. This means that the more finely divided a solid or liquid reactant the greater its surface area per unit volume and the more contact it with the other reactant, thus the faster the reaction. To make an analogy, for example, when one starts a fire, one uses wood chips and small branches — one does not start with large logs right away. In organic chemistry, on water reactions are the exception to the rule that homogeneous reactions take place faster than heterogeneous reactions.

Surface Area of Solids

In a solid, only those particles that are at the surface can be involved in a reaction. Crushing a solid into smaller parts means that more particles are present at the surface, and the frequency of collisions between these and reactant particles increases, and so reaction occurs more rapidly. For example, Sherbet (powder) is a mixture of very fine powder of malic acid (a weak organic acid) and sodium hydrogen carbonate. On contact with the salivain the mouth, these chemicals quickly dissolve and react, releasing carbon dioxide and providing for the fizzy sensation. Also, fireworks manufacturers modify the surface area of solid reactants to control the rate at which the fuels in fireworks are oxidised, using this to create different effects. For example, finely divided aluminium confined in a shell explodes violently. If larger pieces of aluminium are used, the reaction is slower and sparks are seen as pieces of burning metal are ejected.

Concentration

The reactions are due to collisions of reactant species. The frequency with which the molecules or ions collide depends upon their concentrations. The more crowded the molecules are, the more likely they are to collide and react with one another. Thus, an increase in the concentrations of the reactants will usually result in the corresponding increase in the reaction rate, while a decrease in the concentrations will usually have a reverse effect. For example, combustion will occur more rapidly in pure oxygen than in air (21% oxygen).

The rate equation shows the detailed dependence of the reaction rate on the concentrations of reactants and other species present. Different mathematical forms are possible depending on the reaction mechanism. The actual rate equation for a given reaction is determined experimentally and provides information about the reaction mechanism. The mathematical expression of the rate equation is often given by,

$$rate = k\,[A]^x\,[B]^y$$

Here 'x' and 'y' are constants for each reactant, while [A] and [B] are molar concentrations of reactants. Also 'k' is the reaction rate constant which can only be determined experimentally.

Temperature

Temperature usually has a major effect on the rate of a chemical reaction. Molecules at a higher temperature have more thermal energy. Although collision frequency is greater at higher temperatures, this alone contributes only a very small proportion to the increase in rate of reaction. Much more important is the fact that the proportion of reactant molecules with sufficient energy to react (energy greater than activation energy: $E > E_a$) is significantly higher and is explained in detail by the Maxwell–Boltzmann distribution of molecular energies.

The 'rule of thumb' that the rate of chemical reactions doubles for every 10 °C temperature rise is a common misconception. This may have been generalized from the special case of biological systems, where the α(temperature coefficient) is often between 1.5 and 2.5.

A reaction's kinetics can also be studied with a temperature jump approach. This involves using a sharp rise in temperature and observing the relaxation time of the return to equilibrium. A particularly useful form of temperature jump apparatus is a shock tube, which can rapidly jump a gas's temperature by more than 1000 degrees.

Catalysts

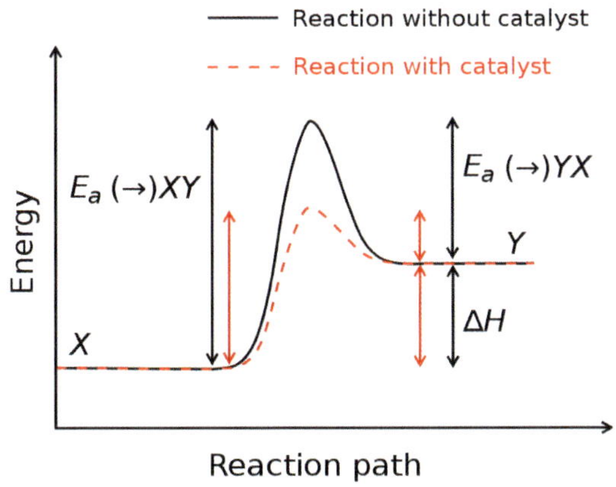

Generic potential energy diagram showing the effect of a catalyst in a hypothetical endothermic chemical reaction. The presence of the catalyst opens a different reaction pathway (shown in red) with a lower activation energy. The final result and the overall thermodynamics are the same

A catalyst is a substance that alters the rate of a chemical reaction but remains chemically unchanged afterwards. The catalyst increases the rate of the reaction by providing a different reaction mechanism to occur with a lower activation energy. In autocatalysis a reaction product is itself a catalyst for that reaction leading to positive feedback. Proteins that act as catalysts in biochemical reactions are called enzymes. Michaelis–Menten kinetics describe the rate of enzyme mediated reactions. A catalyst does not affect the position of the equilibrium, as the catalyst speeds up the backward and forward reactions equally.

In certain organic molecules, specific substituents can have an influence on reaction rate in neighbouring group participation.

Pressure

Increasing the pressure in a gaseous reaction will increase the number of collisions between reactants, increasing the rate of reaction. This is because the activity of a gas is directly proportional to the partial pressure of the gas. This is similar to the effect of increasing the concentration of a solution.

In addition to this straightforward mass-action effect, the rate coefficients themselves can change due to pressure. The rate coefficients and products of many high-temperature gas-phase reactions change if an inert gas is added to the mixture; variations on this effect are called fall-off and chemical activation. These phenomena are due to exothermic or endothermic reactions occurring faster than heat transfer, causing the reacting molecules to have non-thermal energy distributions (non-Boltzmann distribution). Increasing the pressure increases the heat transfer rate between the reacting molecules and the rest of the system, reducing this effect.

Condensed-phase rate coefficients can also be affected by (very high) pressure; this is a completely different effect than fall-off or chemical-activation. It is often studied using diamond anvils.

A reaction's kinetics can also be studied with a pressure jump approach. This involves making fast changes in pressure and observing the relaxation time of the return to equilibrium.

Presence of Light

Light provides necessary activation energy to the starting materials, therefore, most of the reactions becomes faster in the presence of light.

Experimental Methods

The experimental determination of reaction rates involves measuring how the concentrations of reactants or products change over time. For example, the concentration of a reactant can be measured by spectrophotometry at a wavelength where no other reactant or product in the system absorbs light.

For reactions which take at least several minutes, it is possible to start the observations after the reactants have been mixed at the temperature of interest.

Fast Reactions

For faster reactions, the time required to mix the reactants and bring them to a specified temperature may be comparable or longer than the half-life of the reaction. Special methods to start fast reactions without slow mixing step include:

- Stopped flow methods, which can reduce the mixing time to the order of a millisecond.
- Chemical relaxation methods such as temperature jump and pressure jump, in which a pre-mixed system initially at equilibrium is perturbed by rapid heating or depressurization so that it is no longer at equilibrium, and the relaxation back to equilibrium is observed. For example, this method has been used to study the neutralization $H_3O^+ + OH^-$ with a half-life of 1 μs or less under ordinary conditions.
- Flash photolysis, in which a laser pulse produces highly excited species such as free radicals, whose reactions are then studied.

Equilibrium

While chemical kinetics is concerned with the rate of a chemical reaction, thermodynamics determines the extent to which reactions occur. In a reversible reaction, chemical equilibrium is reached when the rates of the forward and reverse reactions are equal [the principle of dynamic equilibrium] and the concentrations of the reactants and Products no longer change. This is demonstrated by, for example, the Haber–Bosch process for combining nitrogen and hydrogen to produce ammonia. Chemical clock reactions such as the Belousov–Zhabotinsky reaction demonstrate that component concentrations can oscillate for a long time before finally attaining the equilibrium.

Free Energy

In general terms, the free energy change (ΔG) of a reaction determines whether a chemical change will take place, but kinetics describes how fast the reaction is. A reaction can be very exothermic and have a very positive entropy change but will not happen in practice if the reaction is too slow. If a reactant can produce two different products, the thermodynamically most stable one will in general form, except in special circumstances when the reaction is said to be under kinetic reaction control. The Curtin–Hammett principle applies when determining the product ratio for two reactants interconverting rapidly, each going to a different product. It is possible to make predictions about reaction rate constants for a reaction from free-energy relationships.

The kinetic isotope effect is the difference in the rate of a chemical reaction when an atom in one of the reactants is replaced by one of its isotopes.

Chemical kinetics provides information on residence time and heat transfer in a chemical reactor in chemical engineering and the molar mass distribution in polymer chemistry.

Applications and Models

The mathematical models that describe chemical reaction kinetics provide chemists and chemical engineers with tools to better understand and describe chemical processes such as food decomposition, microorganism growth, stratospheric ozone decomposition, and the chemistry of biological systems. These models can also be used in the design or modification of chemical reactors to optimize product yield, more efficiently separate products, and eliminate environmentally harmful by-products. When performing catalytic cracking of heavy hydrocarbons into gasoline and light gas, for example, kinetic models can be used to find the temperature and pressure at which the highest yield of heavy hydrocarbons into gasoline will occur.

Chemical Kinetics is frequently validated and explored through modeling in specialized packages as a function of ordinary differential equation-solving (ODE-solving) and curve-fitting.

Catalysts

Catalyst, in chemistry, any substance that increases the rate of a reaction without itself being consumed. Enzymes are naturally occurring catalysts responsible for many essential biochemical reactions.

Most solid catalysts are metals or the oxides, sulfides, and halides of metallic elements and of the semimetallic elements boron, aluminum, and silicon. Gaseous and liquid catalysts are commonly used in their pure form or in combination with suitable carriers or solvents; solid catalysts are commonly dispersed in other substances known as catalyst supports.

In general, catalytic action is a chemical reaction between the catalyst and a reactant, forming chemical intermediates that are able to react more readily with each other or with another reactant, to form the desired end product. During the reaction between the chemical intermediates and the reactants, the catalyst is regenerated. The modes of reactions between the catalysts and the reactants vary widely and in solid catalysts are often complex. Typical of these reactions are acid–base reactions, oxidation–reduction reactions, formation of coordination complexes, and formation of free radicals. With solid catalysts the reaction mechanism is strongly influenced by surface properties and electronic or crystal structures. Certain solid catalysts, called polyfunctional catalysts, are capable of more than one mode of interaction with the reactants; bifunctional catalysts are used extensively for reforming reactions in the petroleum industry.

Catalyzed reactions form the basis of many industrial chemical processes. Catalyst manufacture is itself a rapidly growing industrial process.

Catalytic Processes and their Catalysts

Process	Catalyst
Ammonia synthesis	Iron
Sulfuric acid manufacture	Nitrogen(II) oxide, platinum
Cracking of petroleum	Zeolites
Hydrogenation of unsaturated hydrocarbons	Nickel, platinum, or palladium
Oxidation of hydrocarbons in automobile exhausts	Copper(II) oxide, vanadium(V) oxide, platinum, palladium
Isomerization of n-butane to isobutane	Aluminum chloride, hydrogen chloride

Catalysis

The term *catalysis* was first employed by the great Swedish chemist Jöns Jacob Berzelius in 1835 to correlate a group of observations made by other chemists in the late 18th and early 19th centuries. These included the enhanced conversion of starch to sugar by acids first observed by Gottlieb Sigismund Constantin Kirchhoff; Sir Humphry Davy's observations that platinum hastens the combustion of a variety of gases; the discovery of the stability of hydrogen peroxide in acid solution but its decomposition in the presence of alkali and such metals as manganese, silver, platinum, and gold; and the observation that the oxidation of alcohol to acetic acid is accomplished in the presence of finely divided platinum. The agents promoting these various reactions were termed catalysts, and Berzelius postulated a special unknown catalytic force to be operating in such processes.

In 1834 the English scientist Michael Faraday had examined the power of a platinum plate to accomplish the recombination of gaseous hydrogen and oxygen (the products of electrolysis of water) and the retardation of that recombination by the presence of other gases, such as ethylene and carbon monoxide. Faraday maintained that essential for activity was a perfectly clean metallic surface (at which the retarding gases could compete with the reacting gases and so suppress activity), a concept that would later be shown to be generally important in catalysis.

Many of the primitive technical arts involved unconscious applications of catalysis. The fermentation of wine to acetic acid and the manufacture of soap from fats and alkalies were well known in man's early history. Sulfuric acid prepared by firing mixtures of sulfur and nitre (sodium nitrate) was an early forerunner of the lead chamber process of sulfuric acid manufacture, in which sulfur dioxide oxidation was accelerated by the addition of oxides of nitrogen. (A mechanism for the latter process was suggested by Sir Humphry Davy in 1812 on the basis of experiments carried out by others.)

In 1850 the concept of a velocity of reaction was developed during studies of hydrolysis, or inversion, of cane sugar. The term *inversion* refers to the change in rotation undergone by monochromatic light when it is passed through the reaction system, a parameter that is easily measured, thereby facilitating study of the reaction. It was found that, at any moment, the rate of inversion was proportional to the amount of cane sugar undergoing transformation and that the rate was accelerated by the presence of acids. (Later it was shown that the rate of inversion was directly proportional to the strength of the acid.) This work was in part the precursor of later studies of reaction velocity and the accelerating influence of higher temperature on that velocity by J.H. van't Hoff, Svante Arrhenius, and Wilhelm Ostwald, all of whom played leading roles in the developing science of physical chemistry. Ostwald's work on reaction velocities led him in the 1890s to define catalysts as substances that change the velocity of a given chemical reaction without modification of the energy factors of the reaction.

This statement of Ostwald was a memorable advance since it implied that catalysts do not change the position of equilibrium in a reaction. In 1877 Georges Lemoine had shown that the decomposition of hydriodic acid to hydrogen and iodine reached the same equilibrium point at 350°C (660°F), 19 percent, whether the reaction was carried out rapidly in the presence of platinum sponge or slowly in the gas phase. This observation has an important consequence: a catalyst for the forward process in a reaction is also a catalyst for the reverse reaction. P.E.M. Berthelot, the distinguished French chemist, confirmed this observation in 1879 with liquid systems, when he found that the reaction of organic acids and alcohols, called esterification, is catalyzed by the presence of small amounts of a strong inorganic acid, just as is the reverse process, the hydrolysis of esters (the reaction between an ester and water).

The deliberate application of catalysts to industrial processes was undertaken in the 19th century. P. Phillips, an English chemist, patented the use of platinum to oxidize sulfur dioxide to sulfur trioxide with air. His process was employed for a time but was abandoned because of loss of activity by the platinum catalyst. Poisons in the reactants were subsequently found to be responsible, and the process became a technical success at the turn of the 20th century. In 1871 an industrial process was developed for the oxidation of hydrochloric acid to chlorine in the presence of cupric salts impregnated in clay brick. The chlorine obtained was employed in the manufacture of bleaching powder (a dry substance that releases chlorine on treatment with acid) by reaction with lime. Again, in this reaction, it was observed that the same equilibrium was reached in both directions. Furthermore, it was found that the lower the temperature, the greater the equilibrium content of chlorine; a working temperature of 450°C (840°F) produced the maximum amount of chlorine in a convenient time.

Toward the close of the 19th century, the classic studies of the eminent French chemist Paul Sabatieron the interaction of hydrogen with a wide variety of organic compounds were carried out using various metal catalysts; this research led to the development of a German patent for the hydrogenation of liquid unsaturated fats to solid saturated fats with nickel catalysts. The development of three important German catalytic processes had great impact on industry at the end of the 19th century and in the early decades of the 20th. One was the so-called contact process for producing sulfuric acid catalytically from the sulfur dioxide produced by smelting operations. Another was the catalytic method for the synthetic production of the valuable dyestuff indigo. The third was the catalytic combination of nitrogen and hydrogen for the production of ammonia—the Haber-Bosch process for nitrogen fixation—developed by the chemists Fritz Haber and Carl Bosch.

Classification of Catalysts

Catalysts may be classified generally according to their physical state, their chemical nature, or the nature of the reactions that they catalyze.

Catalysts may be gases, liquids, or solids. In homogeneous catalysis, the catalyst is molecularly dispersed in the same phase (usually gaseous or liquid) as the reactants. In heterogeneous catalysis the reactants and the catalyst are in different phases, separated by a phase boundary. Most commonly, heterogeneous catalysts are solids, and the reactants are gases or liquids.

Homogeneous Catalysis

When the catalyst and the reacting substances are present together in a single state of matter, usually as a gas or a liquid, it is customary to classify the reactions as cases of homogeneous catalysis. Oxides of nitrogen serve as catalysts for the oxidation of sulfur dioxide in the lead chamber process for producing sulfuric acid, an instance of homogeneous catalysis in which the catalyst and reactants are gases. Traces of water vapour catalyze some gas reactions—for example, the interaction of carbon monoxide and oxygen, which proceeds only slowly in dry conditions. Sulfuric acid used as a catalyst for the formation of diethyl ether from ethyl alcohol is an example of homogeneous catalysis in the liquid phase (when the products, water and ether, are continuously removed by distillation); by this method, considerable quantities of alcohol can be converted to ether with a single charge of sulfuric acid. The inversion of cane sugar and the hydrolysis of esters by acid solutions also are examples of homogeneous catalysis in the liquid phase.

The oxidation of sodium sulfite solutions by dissolved oxygen is greatly accelerated by minute traces of copper ions in the homogeneous liquid system. This system is of special interest since it has been shown that the process is a chain reaction. In this case many thousands of molecules of sodium sulfite can be oxidized to sulfate if the initial activation process is produced by absorption of a limited number of quanta (discrete energy measures) of light. The best example of a light-initiated chain reaction is the photocombination of hydrogen (H_2) and chlorine (Cl_2); as many as one million molecules of hydrogen chloride can be formed by absorption of a single light quantum (designated $h\nu$). Here the sequence of reaction is,

$$Cl_2 + h\nu \rightarrow 2Cl\cdot$$

$$Cl\cdot + H_2 \rightarrow HCl + H\cdot$$

$$H\cdot Cl_2 \rightarrow HCl + Cl\cdot$$

with reactions 2 and 3 repeated over and over again. It is of interest to note that such chain reactions can be retarded by the presence of negative catalysts, more commonly termed inhibitors. These are materials that slow down the overall reaction by shortening the reaction chains, generally by entering into a non-chain reaction with one of the chemical components that maintain the chain. A wide variety of substances—including alcohols, sugars, and phenols—have been found to act as inhibitors of the oxidation of sulfite solutions.

A generalized treatment of homogeneous catalysis by acids and bases was given by the Danish physical chemist J.N. Brønsted in the mid-1920s on the basis of his concept of acids and bases. According to Brønsted, an acid is a molecule that can furnish a proton, and a base is a molecule that takes up a proton. On this assumption, the range of acids includes such varied materials as bisulfate ion, HSO_4^- ; acetic acid, CH_3COOH; water, H_2O; hydronium ion, H_3O^+; and ammonium ion, NH_4^+ . The corresponding bases are sulfate ion, SO_4^{2-} ; acetate ion, CH_3COO^-; hydroxide ion, OH^-; water, H_2O; and ammonia, NH_3 (these substances accept protons to yield the listed acids). Brønsted studied a number of acid- and base-catalyzed reactions, including (1) the acid-catalyzed hydrolysis of an ester, ethyl orthoacetate, (2) the basic catalysis of nitramide decomposition ($H_2N{-}NO_2 \rightarrow H_2O + N_2O$), and (3) the acid-base catalysis of the conversion (mutarotation) of glucose to a closely related form. In each case, he observed a direct relationship between the velocity of the catalyzed reaction and the concentration of the catalyzing substance.

Based in part on the ideas of Brønsted, a general scheme for a change of a substance *A* to another substance *B* catalyzed by a material *C* can be formulated thus:

$$(1) \qquad A + C \underset{k}{\overset{k_1}{\rightleftharpoons}} Z$$

$$Z \xrightarrow{K_3} B + C$$

The designation *Z* refers to an intermediate stage, which is formed with a velocity indicated by k_1; *Z* can disappear either to reform $A + C$, with a velocity k_2, or it can decompose by path 2, with velocity k_3, to give the product *B* and regenerate the catalyst *C*. If k_3 is much greater than k_2, the intermediate *Z* is used up almost as quickly as it is formed. The product will then be formed at a rate governed by the expression $k_1[A][C]$, in which the square brackets indicate concentrations of reactant and catalyst. If $k_2$is much larger than k_3, however, the velocity-determining process is the decomposition of *Z*, the rate of formation of the product being represented by the expression $k_3[Z]$, in which $[Z]$ is the concentration of the intermediate. Examples of both types of change have been studied.

In certain instances two or more catalysts present at the same time produce effects greater than either would produce alone. It is then customary to speak of promoter action. Thus, iron ions in solution fortify the action of copper ions in catalyzing a reaction between hydrogen peroxide and iodine. It is assumed that each catalyst activates only one of the reactants.

The most important modern examples of homogeneous catalyses are found in the petrochemical industry. The oxo reaction is one such process: carbon monoxide and hydrogen are added to olefins(unsaturated hydrocarbons) at around 150°C (300°F) and 200 atmospheres of pressure to form aldehydes and alcohols, oxygen-containing organic compounds. A cobalt carbonyl catalyst $Co_2(CO)_8$ is employed; this hydrocarbon-soluble catalyst is believed to activate hydrogen by formation of $HCo(CO)_4$, which then reacts with the olefin. This reaction has led to a number of studies of organometallic chemistry. Copper, silver, and mercury cations (positively charged ions) and permanganate anions (negative ions) also are known to act as homogeneous catalysts for hydrogen activation. Palladium chloride is employed industrially in the catalytic oxidation of ethylene to acetaldehyde in the presence of cupric chloride. The palladium is presumed to be repeatedly converted from the salt to the free metal, the function of the cupric chloride being to participate in the re-formation of the palladium salt from the metal.

Phosphoric, sulfuric, sulfonic, and hydrobromic acids are important agents in the industrial processes of isomerization, polymerization, hydration, and dehydration, as well as in the classic esterification reactions. Free radicals (molecular fragments bearing unpaired electrons) that are generated by the decomposition of peroxides or metal alkyls also initiate homogeneous catalytic processes.

Heterogeneous Catalysis

Many catalytic processes are known in which the catalyst and the reactants are not present in the same phase—that is, state of matter. These are known as heterogeneous catalytic reactions. They include reactions between gases or liquids or both at the surface of a solid catalyst. Since the surface is the place at which the reaction occurs, it generally is prepared in ways that produce large surface areas per unit of catalyst; finely divided metals, metal gauzes, metals incorporated into supporting matrices, and metallic films have all been used in modern heterogeneous catalysis. The metals themselves are used, or they are converted to oxides, sulfides, or halides.

With solid catalysts, at least one of the reactants is chemisorbed (a portmanteau term for chemically adsorbed) by the catalyst. A catalyst is chosen that releases the products formed as readily as possible; otherwise the products remain on the catalyst surface and act as poisons to the process. Chemisorption can occur over a wide temperature range, the most effective temperature for adsorption depending on the nature of the catalyst. Thus, hydrogen is chemisorbed readily by many metals even at liquid air temperatures (below −180°C [−290°F]). With a series of hydrogenation-dehydrogenation catalysts—e.g., zinc oxide–chromic oxide ($ZnO–Cr_2O_3$)—chemisorption of hydrogen often occurs above room temperature. Nitrogen is rapidly chemisorbed on synthetic ammonia-ironcatalyst in the region above 400°C (750°F). It has been shown that iron films chemisorb nitrogen even at liquid air temperatures, with additional chemisorption found above room temperatures. It follows from such considerations that whereas physical adsorptions, which parallel the ease of liquefaction of the adsorbed substance, occur spontaneously, chemisorption, which involves the making and breaking of chemical bonds, often requires activation energies (energy needed to initiate reactions) as do uncatalyzed chemical processes. To be efficient catalytically, a process must involve energies of activation for all the steps involved that, at their maxima, are less than those required for the uncatalyzed reaction. This situation is illustrated graphically in the figure, with a hypotheticalreaction that could occur by either an uncatalyzed or a catalyzed route.

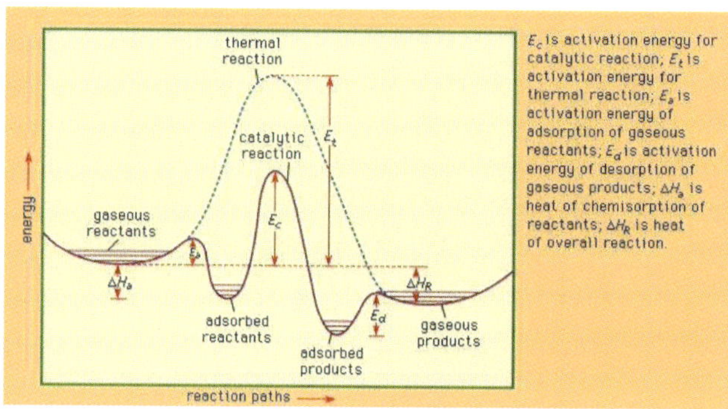

Energy profiles for catalytic and thermal (noncatalytic) reactions in the gaseous phase

Two competing proposals have been made concerning the mechanism of catalytic reactions at surfaces, and it has not been possible to choose between them. Originally, Irving Langmuir, an American physical chemist, proposed chemisorption of both reacting species at the surface, followed by interaction between adjacent species and evaporation of the products. An alternative proposal involves interaction between an impinging molecule and species already adsorbed on the surface. Subsequent developments have suggested various modes of attachment of the adsorbed and adsorbing species.

A major advance in the science of surface catalysis was the development of a method for determining the surface area of catalysts (and other materials) by measuring the multimolecular adsorption of nitrogen at liquid nitrogen temperatures or the adsorption of other gases close to their boiling points. It then became possible to calculate a quantity, designated *Vm*, that represents the volume of gas necessary to form a monolayer on the accessible surface; furthermore, the area of the surface can be determined from the known dimensions of the adsorbed molecules. It has also been found possible to titrate (measure quantitatively) the area of surfaces by chemisorption of gases. Since heterogeneously catalyzed reactions occur on the surface of the catalyst, the rates of such reactions are proportional to the accessible surface area of the catalyst. Active catalysts are thus usually highly porous solids with total surface areas as high as several hundred square metres per gram.

When measurements of surface areas became possible, it was seen at once that many constituents present in minor quantities in the main catalyst material—known as promoters—could act by extending the effective surface area of the catalyst. It also was shown, however, that a promoter might produce an increase in the quality of the surface for the given reaction. Acting in a reverse direction are minor constituents of the reacting system or unwanted products of the reaction, which by preferential adsorption on the reaction sites and resistance to removal give rise to poisons for the process. Poisoning of a catalyst may also result from the poison adversely modifying the electronic properties of the catalyst.

Much can be learned about mechanisms of surface processes by studying the behaviour of isotopic species of the reactants and products on the catalyst. An example of such use concerns the technically important synthesis of ammonia from its elements, the well-known Haber-Bosch process on promoted-iron catalysts. The synthesis of ammonia involves three types of bonds—hydrogen-hydrogen, nitrogen-hydrogen, and nitrogen-nitrogen—all of which can be studied using isotopes of hydrogen and nitrogen. The first of these can be examined on the reduced-iron catalyst by following the progress of the reaction $H_2 + D_2 \rightarrow 2HD$ (in which D equals the deuterium atom, an isotope of hydrogen) on the surface. The reaction is found to occur rapidly even at liquid air temperatures. Nitrogen-hydrogen bond activation can be studied by following the reaction $NH_3 + ND_3 \rightarrow NH_2D + NHD_2$. This proceeds steadily at room temperatures. The reaction involving only N–N bonds, however, studied by following the process $^{14}N_2 + {}^{15}N_2 \rightarrow 2{}^{14}N{-}{}^{15}N$ (in which ^{14}N and ^{15}N are stable isotopes of nitrogen), is shown to proceed only at the higher temperatures of ammonia synthesis, around 400°C (750°F). From these data one concludes that the activation of the nitrogen molecules is the slow step (the process that limits the overall reaction) in ammonia synthesis. This conclusion is confirmed by measurement of rate of adsorption of nitrogen on the iron catalysts. Other, similar isotopic studies have yielded valuable information on the reactions of hydrocarbons, using deuterium and carbon-14 as the isotopic tracers.

Photocatalysts

Photocatalysis is the phenomenon where the catalyst can receive light (such as visible light), be promoted to an excited state, and then undergo intersystem crossing with the starting material, returning to ground state without being consumed. The excited state of the starting material will then undergo reactions it ordinarily could not if directly illuminated. For example, singlet oxygen is usually produced by photocatalysis. Photocatalysts are also the main ingredient in dye-sensitized solar cells.

Enzymes and Biocatalysts

In biology, enzymes are protein-based catalysts in metabolism and catabolism. Most biocatalysts are enzymes, but other non-protein-based classes of biomolecules also exhibit catalytic properties including ribozymes, and synthetic deoxyribozymes.

Biocatalysts can be thought of as intermediate between homogeneous and heterogeneous catalysts, although strictly speaking soluble enzymes are homogeneous catalysts and membrane-bound enzymes are heterogeneous. Several factors affect the activity of enzymes (and other catalysts) including temperature, pH, concentration of enzyme, substrate, and products. A particularly important reagent in enzymatic reactions is water, which is the product of many bond-forming reactions and a reactant in many bond-breaking processes.

In biocatalysis, enzymes are employed to prepare many commodity chemicals including high-fructose corn syrup and acrylamide.

Some monoclonal antibodies whose binding target is a stable molecule which resembles the transition state of a chemical reaction can function as weak catalysts for that chemical reaction by lowering its activation energy. Such catalytic antibodies are sometimes called "abzymes".

Nanocatalysts

Nanocatalysts are nanomaterials with catalytic activities. They have been extensively explored for wide range of applications. Among them, the nanocatalysts with enzyme mimicking activities are collectively called as nanozymes.

Tandem Catalysis

In tandem catalysis two or more different catalysts are coupled in a one-pot reaction.

Autocatalysis

In autocatalysis the catalyst *is* a product of the overall reaction, in contrast to all other types of catalysis considered in this article. The simplest example of autocatalysis is a reaction of type $A + B \rightarrow 2B$, in one or in several steps. The overall reaction is just $A \rightarrow B$, so that B is a product. But since B is also a reactant, it may be present in the rate equation and affect the reaction rate. As the reaction proceeds, the concentration of B increases and can accelerate the reaction as a catalyst. In effect, the reaction accelerates itself or is autocatalyzed.

A real example is the hydrolysis of an ester such as aspirin to a carboxylic acid and an alcohol. In the absence of added acid catalysts, the carboxylic acid product catalyzes the hydrolysis.

General Principles

Units

Catalytic activity is usually denoted by the symbol z and measured in mol/s, a unit which was called katal and defined the SI unit for catalytic activity since 1999. Catalytic activity is not a kind of reaction rate, but a property of the catalyst under certain conditions, in relation to a specific chemical reaction. Catalytic activity of one katal (Symbol 1 kat = 1 mol/s) of a catalyst means one mole of that catalyst (substance, in Mol) will catalyse 1 mole of the reactant to product in one second. A catalyst may and usually will have different catalytic activity for distinct reactions.

Typical Mechanism

Catalysts generally react with one or more reactants to form intermediates that subsequently give the final reaction product, in the process regenerating the catalyst. The following is a typical reaction scheme, where *C* represents the catalyst, X and Y are reactants, and Z is the product of the reaction of X and Y:

$$X + C \rightarrow XC$$

$$Y + XC \rightarrow XYC$$

$$XYC \rightarrow CZ$$

$$CZ \rightarrow C + Z$$

Although the catalyst is consumed by first reaction, it is subsequently produced by forth reaction, so it does not occur in the overall reaction equation:

$$X + Y \rightarrow Z$$

As a catalyst is regenerated in a reaction, often only small amounts are needed to increase the rate of the reaction. In practice, however, catalysts are sometimes consumed in secondary processes.

The catalyst does usually appear in the rate equation. For example, if the rate-determining step in the above reaction scheme is the first step X + C → XC, the catalyzed reaction will be second order with rate equation $v = k_{cat}[X][C]$, which is proportional to the catalyst concentration [C]. However [C] remains constant during the reaction so that the catalyzed reaction is pseudo-first order: $v = k_{obs}\left[X\right]$, *where* $k_{obs} = k_{cat}\left[C\right]$.

As an example of a detailed mechanism at the microscopic level, in 2008 Danish researchers first revealed the sequence of events when oxygen and hydrogen combine on the surface of titanium dioxide (TiO_2, or *titania*) to produce water. With a time-lapse series of scanning tunneling microscopy images, they determined the molecules undergo adsorption, dissociation and diffusion before reacting. The intermediate reaction states were: HO_2, H_2O_2, then $H_3O_2$and the final reaction product (water molecule dimers), after which the water molecule desorbs from the catalyst surface.

Reaction Energetics

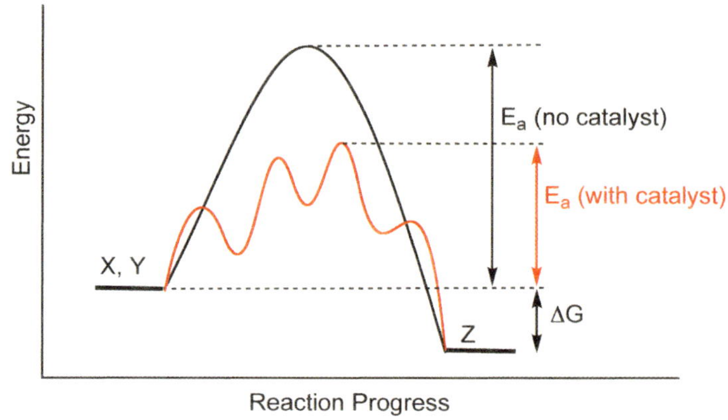

Generic potential energy diagram showing the effect of a catalyst in a hypothetical exothermic chemical reaction X + Y to give Z. The presence of the catalyst opens a different reaction pathway (shown in red) with a lower activation energy. The final result and the overall thermodynamics are the same

Catalysts work by providing an (alternative) mechanism involving a different transition state and lower activation energy. Consequently, more molecular collisions have the energy needed to reach the transition state. Hence, catalysts can enable reactions that would otherwise be blocked or slowed by a kinetic barrier. The catalyst may increase reaction rate or selectivity, or enable the reaction at lower temperatures. This effect can be illustrated with an energy profile diagram.

In the catalyzed elementary reaction, catalysts do not change the extent of a reaction: they have no effect on the chemical equilibrium of a reaction because the rate of both the forward and the reverse reaction are both affected. The second law of thermodynamics describes why a catalyst does not change the chemical equilibrium of a reaction. Suppose there was such a catalyst that shifted an equilibrium. Introducing the catalyst to the system would result in a reaction to move to the new equilibrium, producing energy. Production of energy is a necessary result since reactions are spontaneous only if Gibbs free energy is produced, and if there is no energy barrier, there is no need for a catalyst. Then, removing the catalyst would also result in reaction, producing energy; i.e. the addition and its reverse process, removal, would both produce energy. Thus, a catalyst that could change the equilibrium would be a perpetual motion machine, a contradiction to the laws of thermodynamics.

If a catalyst does change the equilibrium, then it must be consumed as the reaction proceeds, and thus it is also a reactant. Illustrative is the base-catalysed hydrolysis of esters, where the produced carboxylic acid immediately reacts with the base catalyst and thus the reaction equilibrium is shifted towards hydrolysis.

The SI derived unit for measuring the catalytic activity of a catalyst is the katal, which is moles per second. The productivity of a catalyst can be described by the turnover number(or TON) and the catalytic activity by the *turn over frequency* (TOF), which is the TON per time unit. The biochemical equivalent is the enzyme unit. For more information on the efficiency of enzymatic catalysis.

The catalyst stabilizes the transition state more than it stabilizes the starting material. It decreases the kinetic barrier by decreasing the *difference* in energy between starting material and transition state. It does not change the energy difference between starting materials and products (thermodynamic barrier), or the available energy (this is provided by the environment as heat or light).

Materials

The chemical nature of catalysts is as diverse as catalysis itself, although some generalizations can be made. Proton acids are probably the most widely used catalysts, especially for the many reactions involving water, including hydrolysis and its reverse. Multifunctional solids often are catalytically active, e.g. zeolites, alumina, higher-order oxides, graphitic carbon, nanoparticles, nanodots, and facets of bulk materials. Transition metals are often used to catalyze redox reactions (oxidation, hydrogenation). Examples are nickel, such as Raney nickel for hydrogenation, and vanadium(V) oxide for oxidation of sulfur dioxide into sulfur trioxide by the so-called contact process. Many catalytic processes, especially those used in organic synthesis, require "late transition metals", such as palladium, platinum, gold, ruthenium, rhodium, or iridium.

Some so-called catalysts are really precatalysts. Precatalysts convert to catalysts in the reaction. For example, Wilkinson's catalyst $RhCl(PPh_3)_3$ loses one triphenylphosphine ligand before entering the true catalytic cycle. Precatalysts are easier to store but are easily activated in situ. Because of this preactivation step, many catalytic reactions involve an induction period.

Chemical species that improve catalytic activity are called co-catalysts (cocatalysts) or promotors in cooperative catalysis.

Significance

Left: Partially caramelised cube sugar, Right: burning cube sugar with ash as catalyst

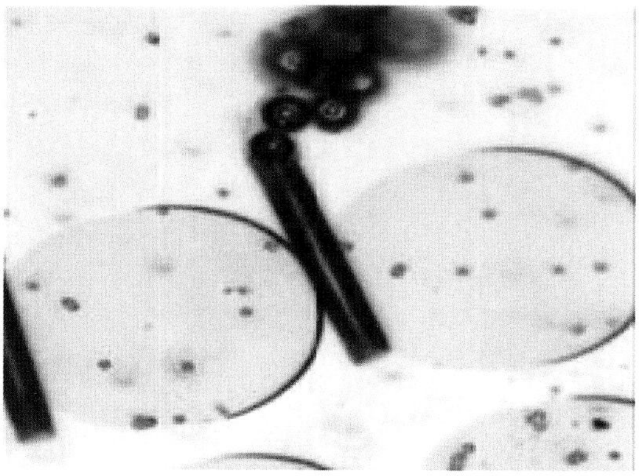

A Ti-Cr-Pt tube (~40 μm long) releases oxygen bubbles when immersed in hydrogen peroxide (via catalytic decomposition), forming a micropump

Estimates are that 90% of all commercially produced chemical products involve catalysts at some stage in the process of their manufacture. In 2005, catalytic processes generated about $900 billion in products worldwide. Catalysis is so pervasive that subareas are not readily classified. Some areas of particular concentration are surveyed below.

Energy Processing

Petroleum refining makes intensive use of catalysis for alkylation, catalytic cracking (breaking long-chain hydrocarbons into smaller pieces), naphtha reforming and steam reforming (conversion of hydrocarbons into synthesis gas). Even the exhaust from the burning of fossil fuels is treated via catalysis: Catalytic converters, typically composed of platinum and rhodium, break down some of the more harmful byproducts of automobile exhaust.

$$2\,CO + 2\,NO \rightarrow 2\,CO_2 + N_2$$

With regard to synthetic fuels, an old but still important process is the Fischer-Tropsch synthesis of hydrocarbons from synthesis gas, which itself is processed via water-gas shift reactions, catalysed by iron. Biodiesel and related biofuels require processing via both inorganic and biocatalysts.

Fuel cells rely on catalysts for both the anodic and cathodic reactions. Catalytic heaters generate flameless heat from a supply of combustible fuel.

Bulk Chemicals

Some of the largest-scale chemicals are produced via catalytic oxidation, often using oxygen. Examples include nitric acid (from ammonia), sulfuric acid (from sulfur dioxide to sulfur trioxide by the contact process), terephthalic acid from p-xylene, and acrylonitrile from propane and ammonia.

Many other chemical products are generated by large-scale reduction, often via hydrogenation. The largest-scale example is ammonia, which is prepared via the Haber process from nitrogen.Methanol is prepared from carbon monoxide.

Bulk polymers derived from ethylene and propylene are often prepared via Ziegler-Natta catalysis. Polyesters, polyamides, and isocyanates are derived via acid-base catalysis.

Most carbonylation processes require metal catalysts, examples include the Monsanto acetic acid process and hydroformylation.

Fine Chemicals

Many fine chemicals are prepared via catalysis; methods include those of heavy industry as well as more specialized processes that would be prohibitively expensive on a large scale. Examples include the Heck reaction, and Friedel–Crafts reactions.

Because most bioactive compounds are chiral, many pharmaceuticals are produced by enantioselective catalysis (catalytic asymmetric synthesis).

Food Processing

One of the most obvious applications of catalysis is the hydrogenation (reaction with hydrogen gas) of fats using nickel catalyst to produce margarine. Many other foodstuffs are prepared via biocatalysis.

Environment

Catalysis impacts the environment by increasing the efficiency of industrial processes, but catalysis also plays a direct role in the environment. A notable example is the catalytic role of chlorine free radicals in the breakdown of ozone. These radicals are formed by the action of ultraviolet radiation on chlorofluorocarbons (CFCs).

$$Cl^{\cdot} + O_3 \rightarrow ClO^{\cdot} + O_2$$

$$ClO^{\cdot} + O^{\cdot} \rightarrow Cl^{\cdot} + O_2$$

Inhibitors, Poisons, and Promoters

Substances that reduce the action of catalysts are called catalyst inhibitors if reversible, and catalyst poisons if irreversible. Promoters are substances that increase the catalytic activity, even though they are not catalysts by themselves.

Inhibitors are sometimes referred to as "negative catalysts" since they decrease the reaction rate. However the term inhibitor is preferred since they do not work by introducing a reaction path with higher activation energy; this would not reduce the rate since the reaction would continue to occur by the non-catalyzed path. Instead they act either by deactivating catalysts, or by removing reaction intermediates such as free radicals.

The inhibitor may modify selectivity in addition to rate. For instance, in the reduction of alkynes to alkenes, a palladium (Pd) catalyst partly "poisoned" with lead(II) acetate ($Pb(CH_3CO_2)_2$) can be used. Without the deactivation of the catalyst, the alkene produced would be further reduced to alkane.

The inhibitor can produce this effect by, e.g., selectively poisoning only certain types of active sites. Another mechanism is the modification of surface geometry. For instance, in hydrogenation operations, large planes of metal surface function as sites of hydrogenolysis catalysis while sites catalyzing hydrogenation of unsaturates are smaller. Thus, a poison that covers surface randomly will tend to reduce the number of uncontaminated large planes but leave proportionally more smaller sites free, thus changing the hydrogenation vs. hydrogenolysis selectivity. Many other mechanisms are also possible.

Promoters can cover up surface to prevent production of a mat of coke, or even actively remove such material (e.g., rhenium on platinum in platforming). They can aid the dispersion of the catalytic material or bind to reagents .

Reaction Mechanism

Reaction mechanism, in chemical reactions, the detailed processes by which chemical substances are transformed into other substances. The reactions themselves may involve the interactions of atoms, molecules, ions, electrons, and free radicals, and they may take place in gases, liquids, or solids—or at interfaces between any of these.

The study of the detailed processes of reaction mechanisms is important for many reasons, including the help it gives in understanding and controlling chemical reactions. Many reactions of great commercial importance can proceed by more than one reaction path; knowledge of the reaction mechanisms involved may make it possible to choose reaction conditions favouring one path over another, thereby giving maximum amounts of desired products and minimum amounts of undesired products. Furthermore, on the basis of reaction mechanisms, it is sometimes possible to find correlations between systems not otherwise obviously related. The ability to draw such analogiesfrequently makes it possible to predict the course of untried reactions. Finally, detailed information about reaction mechanisms permits unification and understanding of large bodies of otherwise unrelated phenomena, a matter of great importance in the theory and practice of chemistry.

Generally, the chemical reactions whose mechanisms are of interest to chemists are those that occur in solution and involve the breaking and reforming of covalent bonds between atoms—covalent bonds being those in which electrons are shared between atoms. Interest in these reactions is especially great because they are the reactions by which such materials as plastics, dyes, synthetic fibres, and medicinal agents are prepared and because most of the biochemical reactions of living systems are of this type.

In addition, reactions of this kind generally occur in timescales convenient for study, neither too fast nor too slow, and under conditions that are easily manipulated for experimental purposes. There are a number of techniques by which the mechanisms of such reactions can be investigated.

Chemical reactions involve changes in bonding patterns of molecules—that is, changes in the relative positions of atoms in and among molecules, as well as shifts in the electrons that hold the atoms together in chemical bonds. Reaction mechanisms, therefore, must include descriptions of these movements with regard to spatial change and also with regard to time. The overall route of change is called the course of the reaction, and the detailed process by which the change occurs is referred to as the reaction path or pathway.

Also important to the study of reaction mechanisms are the energy requirements of the reactions. Most reactions of mechanistic interest are activated processes—that is, processes that must have a supply of energy before they can occur. The energy is consumed in carrying the starting material of the reaction over an energy barrier. This process occurs when the starting material absorbs energy and is converted to an activated complex or transition state. The activated complex then proceeds to furnish the product of the reaction without further input of energy—often, in fact, with a release of energy. Such considerations are important to an understanding of reaction mechanisms because the actual course that any reaction follows is the one that requires the least energy of activation. This reaction course is not always the one that would seem simplest to the chemist without detailed study of the different possible mechanisms.

The study of reaction mechanisms is complicated by the reversibility of most reactions (the tendency of the reaction products to revert to the starting materials) and by the existence of competing reactions (reactions that convert the starting material to something other than the desired products). Another complicating factor is the fact that many reactions occur in stages in which intermediate products (intermediates) are formed and then converted by further reactions to the final products. In examining chemical reactions, it is useful to consider several general subjects: (1) factors that influence the course of chemical reactions, (2) energy changes involved in the course of a typical reaction, (3) factors that reveal the mechanism of a reaction, and (4) the classification of reaction mechanisms. With this information in mind, it is then possible to look briefly at some of the more important classes of reaction mechanisms.

General Considerations

Determinants of the Course of Reaction

The Reactants

In analyzing the mechanism of a reaction, account must be taken of all the factors that influence its course. After the bulk chemical constituents have been identified by ordinary methods of structure determination and analysis, any prereaction changes involving the reactants, either individually or together, must be investigated. Thus, in the cleavage of the substance ethyl acetate by water(hydrolysis), the actual reagent that attacks the ethyl acetate molecule may be the water molecule itself, or it may be the hydroxide ion (OH–) produced from it.

The hydrolysis of ethyl acetate can be represented by the following equation:

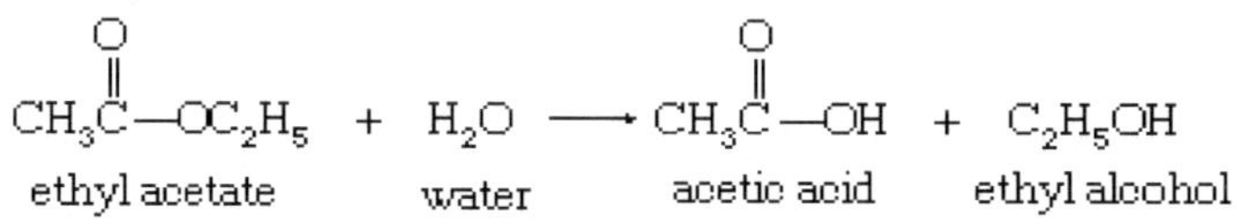

In which the structures of the molecules are represented schematically by their structural formulas. An arrow is used to indicate the reaction, with the formulas for the starting materials on the left and those of the products on the right. In the structural formulas, the atoms of the elements are represented by their chemical symbols (C for carbon, H for hydrogen, and O for oxygen), and the numbers of the atoms in particular groups are designated by numeral subscripts. The chemical bonds of greatest interest are represented by short lines between the symbols of the atoms connected by the bonds.

Important to this reaction is an equilibrium involving the cleavage of the water molecules into positively and negatively charged particles (ions), as follows:

$$2H_2O \rightleftharpoons H_3O^+ + OH^-$$

water hydroniumion hydroxideion

In this equation the numeral in front of the symbol for the water molecule indicates the number of molecules involved in the reaction. The composite arrow indicates that the reaction can proceed in either direction, starting material being converted to products and vice versa. In practice, both reactions occur together, and a balance, or equilibrium, of starting materials and products is set up. The significance of this equilibrium for the hydrolysis of ethyl acetate is that any of the three entities (water molecules, hydronium, or hydroxide ions) may be involved in the reaction, and the mechanism is not known until it is established which of these is the actual participant. This often can be established if it is possible to determine the relative amounts of the three in the reaction medium and if it can be shown that the rate of the reaction depends upon the amount (or concentration) of one of them. Under certain conditions the hydrolysis of ethyl acetate is found to involve water molecules (as shown in the equation above); in other cases, hydroxide ion is involved.

The Transition State

The transition state, or activated complex, is the fleeting molecular configuration that exists at the top of the energy barrier that the reactants must surmount to become the products. It is not strictly a component of the reaction system, and it cannot be examined directly in the way that an intermediate (however unstable) can, because it lasts no longer than the duration of a molecular collision. The transition state may have properties of its own, not reflected in those of the starting materials or of the products and of the reaction, and so it is of vital importance in determining the course of reaction. Inference concerning the nature of the transition state is the essence of mechanistic study.

The Solvent

The solvent, or medium in which the reaction occurs, may perform the mechanical—but often vital—role of allowing otherwise immiscible reactants to come together rapidly. Among the important groups of solvents, each with its own special type of behaviour, are hydroxylic solvents (the molecules of which contain hydroxyl [–OH] groups, such as water and alcohols), dipolar aprotic solvents (the molecules of which show a separation of electrical charge but do not easily give up a proton, or positive hydrogen ion; e.g., acetone), and nonpolar solvents (the molecules of which do not show charge separation; e.g., hexane).

In dissolving the reactants, the solvent may interact with any or all of them, and it may be involved in the transition state for any reaction available for the system. If the solvent interacts more powerfully with the transition state than with the reactants, it facilitates the reaction. The solvent itself may of course be one of the reactants, and this circumstance introduces special problems because of the difficulty of distinguishing experimentally between its functions as a reagent and as an environment for the reaction.

Catalysts

Catalysts are substances that speed up a reaction by facilitating a particular mechanism—sometimes by influencing an existing prereaction and sometimes by making a new process energetically favourable. Their presence or absence frequently determines the course a reaction may take, simply because one of a number of competing reactions is, or is not, favoured. (Most catalysts are changed chemically while they speed up a reaction; sometimes—but not always—they are consumed, and sometimes they are reformed and so appear to be unchanged in concentration during a reaction.)

The Products

All reactions are reversible in principle, and the nature of the products of the reaction can affect the reaction course in a number of ways. When the position of equilibrium is unfavourable, for example, the accumulation of products may cause a reversal of the reaction. In such circumstances, the physical removal of the products (through their volatility or insolubility, for example) facilitates the completion of the forward process.

Sometimes too, one of the products acts as a catalyst or as an inhibitor, behaviour that strongly influences the course of the reaction.

The Reaction Conditions

The conditions under which some reaction occurs, including such variables as the temperature and concentrations of reactants, also are important in determining the course of the reaction. For reactions that have a high energy barrier between reactants and products, the rate is highly responsive to change in temperature, and such reactions become more likely at increased temperatures, so the minor products of a reaction often appear in larger proportions at higher temperatures.

Similarly, the concentration of reagents can be important to the course of a reaction, especially if two mechanisms are available that involve different numbers of molecules in the transition states. Higher concentrations of a particular reagent favour those mechanisms in which greater numbers of molecules are involved in the transition state. The pressure applied to the reacting system also may be significant, partly because it has an effect on concentration and partly because mechanisms involving closely associated transition states become more favourable at high pressures. The latter relationship comes about because associated transition states are those in which several molecules or ions are brought close together (and therefore take up less space), a situation that is encouraged by increased pressures.

Energy Changes Involved in Reactions

Collisions between molecules are rapid; therefore, reactions that occur spontaneously whenever the reagents collide are fast at ordinary concentrations. However, a reaction may be restricted in rate by its dependence on the occurrence of molecular collisions, because, for example, the reagents are present in such small amounts that reactions can only occur when they happen to encounter one another. Such a reaction is said to be diffusion-controlled, because it is dependent on the process of diffusion to bring the molecules together. In such cases the viscosity of the medium is relevant; the more viscous, or "thick", the medium, the more difficult the diffusion and the slower the reaction.

However, most reactions involve a rate-limiting energy barrier, and it is the nature of this barrier and of the molecular configuration at its top that determines the mechanism. Diagrams of energy changes during the course of reaction often are used to illustrate the energetic aspects of the reaction. An example of a possible energy diagram for a hypothetical one-stage process—the dissociation in a solution of a covalent molecule designated E–N, into its ions, E^+ and N^-.

In this diagram the energy is plotted against a reaction coordinate—a spatial relationship that varies smoothly during the course of the reaction—in this case the distance between the portions of the molecule designated E and N. At the left-hand energy minimum of the figure, the bond between E and N is fully formed; if energy is applied to excite the system in such a way that E and N are brought even closer together (region *a* of the figure), the atomic nuclei repel, with the result that energy rises steeply. Alternatively, excitation energy, such as thermal energy from collisions with other molecules, may stretch the bond, and the energy curve then moves into region *b*. The E–N bond is thereby weakened steadily until the transition state is reached. This point, as can be seen in the figure, has an energy peak on the reaction coordinate. At the same time, this point represents the minimum energy required to convert the reactants into the products. The curve shown should be considered as only a planar section of a three-dimensional energy surface relating to the various possible spatial relations between the components of the reaction. The passage of the reactants from the initial state to the products then can be thought of as analogous to the climb from a valley (the initial state) through the lowest mountain pass leading to a second valley (the products). Thus, although the transition state represents a peak on the single curve depicted, it really represents a secondary minimum (or pass) in the energy surface. From the top of this pass (the transition state), the molecule can only descend, losing energy by collision. In doing so it may revert to the starting materials, or it may dissociate to give the products (region *c* in the figure). The products in the case of the reaction chosen are the ions E^+ and N^-, which are held together by electrostatic attraction as an ion pair at the right-hand minimum of the graph; beyond this point, further separation of the ions involves the consumption of energy (region *d*). In principle, the products may lie at higher or lower energy than that of the initial state. The mechanism of a reaction such as this may be considered to be completely defined when the structures and energy properties of the starting materials, the products, and the transition state are known.

Ultimately, it should become possible to compute the properties of the molecules solely from the properties of their constituent atoms and also to deduce the transition states for any of the reactions these molecules may undergo. For a few simple situations, approaches already have been made to definitions of mechanism in this degree of detail. Systems involving several atoms, however, require a many-dimensional representation of the reaction course instead of the two-dimensional description shown in the figure. The problems of computation, and of testing theory against experiment, then become enormous. Nonetheless, attempts have been made to deal with some simple reactions of systems involving up to about five atoms.

For a reaction involving several distinct stages, a more complicated description of the reaction course also is necessary. The figure gives an example of such a situation. This hypothetical reaction is reversible, with three successive intermediate complexes formed between the reactants. Unlike the case of the simpler situation above, the physical process that best approximates what is happening along the reaction coordinate changes from stage to stage across the diagram.

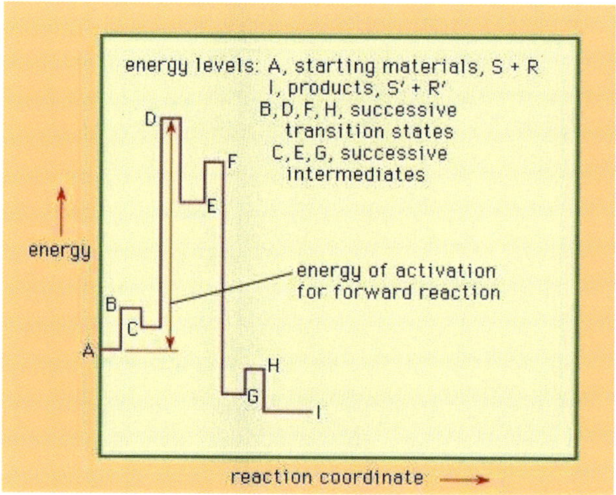

Energy levels in a hypothetical multistage reaction

The highest point on the energy diagram corresponds to the energy of the transition state of the rate-limiting step in the reaction—that is, the slowest step in the reaction, the one that governs, or limits, the overall rate. The rate of reaction is independent of the nature and number of the intermediates that lie before this transition state on the reaction coordinate. Progress along the reaction coordinate cannot be identified with the time course of the reaction because any individual pair of reactants may reside for an appreciable time in the partially activated state represented by one of the intermediate complexes before final reaction is achieved. Once the reactants have passed the rate-limiting transition state, they must lose energy (usually by collision with other molecules) to reach the final state. In a sense, the reaction coordinate in such a reaction may be thought of as representing the chemical course of the reaction (rather than a spatial or a time course). If the diagram represents the only reaction of the system, then it is possible to apply the principle of microscopic reversibility, which states that the course taken by reverse action will be statistically identical with that taken by the forward reaction. This principle is not applicable to a reaction giving several different products not at equilibrium with one another, however.

Reaction Intermediates

Reaction intermediates are chemical species, often unstable and short-lived (however sometimes can be isolated), which are not reactants or products of the overall chemical reaction, but are temporary products and/or reactants in the mechanism's reaction steps. Reaction intermediates are often free radicals or ions.

The kinetics (relative rates of the reaction steps and the rate equation for the overall reaction) are explained in terms of the energy needed for the conversion of the reactants to the proposed transition states (molecular states that corresponds to maxima on the reaction coordinates, and to saddle points on the potential energy surface for the reaction).

Chemical Kinetics

Information about the mechanism of a reaction is often provided by the use of chemical kinetics to determine the rate equation and the reaction order in each reactant.

Consider the following reaction for example:

$$CO + NO_2 \rightarrow CO_2 + NO$$

In this case, experiments have determined that this reaction takes place according to the rate law $r = k[NO_2]^2$. This form suggests that the rate-determining step is a reaction between two molecules of NO_2. A possible mechanism for the overall reaction that explains the rate law is:

$$2\,NO_2 \rightarrow NO_3 + NO\ (slow)$$

$$NO_3 + CO \rightarrow NO_2 + CO_2\ (fast)$$

Each step is called an elementary step, and each has its own rate law and molecularity. The elementary steps should add up to the original reaction. (Meaning, if we were to cancel out all the molecules that appear on both sides of the reaction, we would be left with the original reaction.)

When determining the overall rate law for a reaction, the slowest step is the step that determines the reaction rate. Because the first step (in the above reaction) is the slowest step, it is the rate-determining step. Because it involves the collision of two NO_2 molecules, it is a bimolecular reaction with a rate law of $r = k[NO_2]^2$.

Other reactions may have mechanisms of several consecutive steps. In organic chemistry, the reaction mechanism for the benzoin condensation, put forward in 1903 by A. J. Lapworth, was one of the first proposed reaction mechanisms.

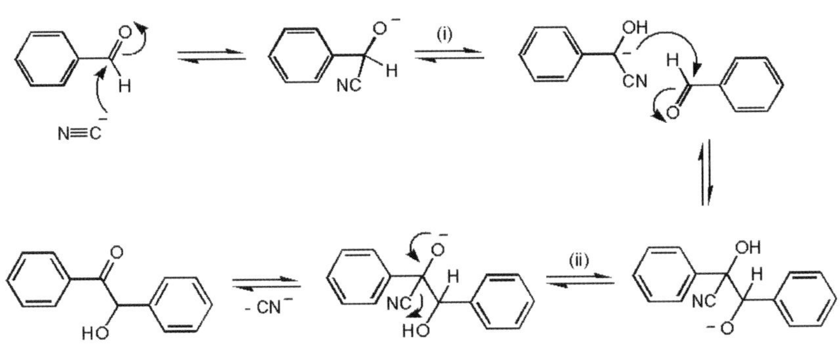

Benzoin condensation reaction mechanism. Cyanide ion (CN^-) acts as a catalyst here, entering at the first step and leaving in the last step. Proton (H^+) transfers occur at (i) and (ii). The arrow pushing method is used in some of the steps to show where electron pairs go

A chain reaction is an example of a complex mechanism, in which the propagation steps form a closed cycle.

Other Experimental Methods to Determine Mechanism

Many experiments that suggest the possible sequence of steps in a reaction mechanism have been designed, including:

- Measurement of the effect of temperature (Arrhenius equation) to determine the activation energy.
- Spectroscopic observation of reaction intermediates.
- Determination of the stereochemistry of products, for example in nucleophilic substitution reactions.
- Measurement of the effect of isotopic substitution on the reaction rate.
- For reactions in solution, measurement of the effect of pressure on the reaction rate to determine the volume change on formation of the activated complex.
- For reactions of ions in solution, measurement of the effect of ionic strength on the reaction rate.
- Direct observation of the activated complex by pump-probe spectroscopy.
- Infrared chemiluminescence to detect vibrational excitation in the products.
- Electrospray ionization mass spectrometry.
- Crossover experiments.

Theoretical Modeling

A correct reaction mechanism is an important part of accurate predictive modeling. For many combustion and plasma systems, detailed mechanisms are not available or require development.

Even when information is available, identifying and assembling the relevant data from a variety of sources, reconciling discrepant values and extrapolating to different conditions can be a difficult process without expert help. Rate constants or thermochemical data are often not available in the literature, so computational chemistry techniques or group additivity methods must be used to obtain the required parameters.

Computational chemistry methods can also be used to calculate potential energy surfaces for reactions and determine probable mechanisms.

Molecularity

Molecularity in chemistry is the number of colliding molecular entities that are involved in a single reaction step.

- A reaction step involving one molecular entity is called unimolecular.

- A reaction step involving two molecular entities is called bimolecular.
- A reaction step involving three molecular entities is called trimolecular or termolecular.

In general, reaction steps involving more than three molecular entities do not occur, because is statistically improbable in terms of Maxwell distribution to find such transition state.

Reaction Rate

Reaction rate, the speed at which a chemical reaction proceeds. It is often expressed in terms of either the concentration (amount per unit volume) of a product that is formed in a unit of time or the concentration of a reactant that is consumed in a unit of time. Alternatively, it may be defined in terms of the amounts of the reactants consumed or products formed in a unit of time. For example, suppose that the balanced chemical equation for a reaction is of the form,

$$A + 3B \rightarrow 2Z.$$

The rate could be expressed in the following alternative ways:

$$d[Z]/dt,\ -d[A]/dt,\ -d[B]/dt, dz/dt,\ -da/dt,\ -db/dt$$

where t is the time, [A], [B], and [Z] are the concentrations of the substances, and a, b, and z are their amounts. Note that these six expressions are all different from one another but are simply related. Chemical reactions proceed at vastly different speeds depending on the nature of the reacting substances, the type of chemical transformation, the temperature, and other factors. In general, reactions in which atoms or ions (electrically charged particles) combine occur very rapidly, while those in which covalent bonds (bonds in which atoms share electrons) are broken are much slower. For a given reaction, the speed of the reaction will vary with the temperature, the pressure, and the amounts of reactants present. Reactions usually slow down as time goes on because of the depletion of the reactants. In some cases the addition of a substance that is not itself a reactant, called a catalyst, accelerates a reaction. The rate constant, or the specific rate constant, is the proportionality constant in the equation that expresses the relationship between the rate of a chemical reaction and the concentrations of the reacting substances. The measurement and interpretation of reactions constitute the branch of chemistry known as chemical kinetics.

The reaction rate can be determined by measuring how fast the concentration of A or B decreases, or by how fast the concentration of AB increases.

$$A + B \rightarrow AB$$

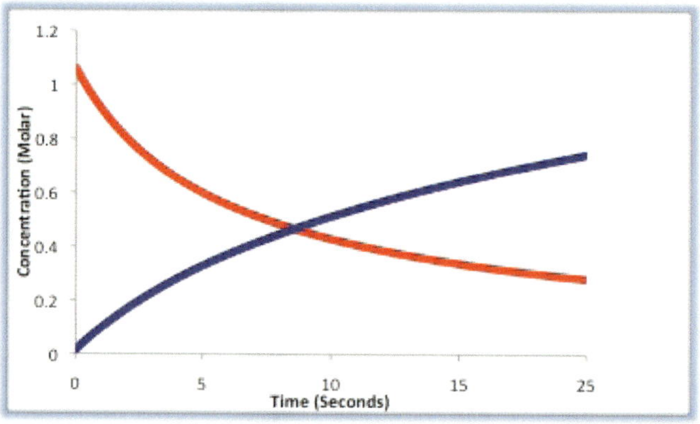

The above picture shows a hypthetical reaction profile in which the reactants (red) decrease in concentration as the products increase in concetration (blue)

For the stochiometrically complicated reaction:

$$aA + bB \rightarrow cC + dD$$

$$Rate = \frac{-1}{a}\frac{d[A]}{dt} = \frac{-1}{b}\frac{d[B]}{dt} = \frac{1}{c}\frac{d[C]}{dt} = \frac{1}{d}\frac{d[D]}{dt}$$

Looking at figure above, we can see that the rate can be measured in terms of either reactant (A or B) or either product (C or D). Not all variables are needed to solve for the rate. Therefore, if you have the value for "A" as well as the value for "a" you can solve for the reaction rate.

You can also notice from formula above that, the change in reactants over the change in time must have a negative sign in front of them. The reason for this is because the reactants are decreasing as a function of time, the rate would come out to be negative (because it is the reverse rate). Therefore, putting a negative sign in front of the variable will allow for the solution to be a positive rate.

Influencing Factors

- The nature of the reaction: Some reactions are naturally faster than others. The number of reacting species, their physical state (the particles that form solids move much more slowly than those of gases or those in solution), the complexity of the reaction and other factors can greatly influence the rate of a reaction.
- Concentration: Reaction rate increases with concentration, as described by the rate law and explained by collision theory. As reactant concentration increases, the frequency of collision increases.
- Pressure: The rate of gaseous reactions increases with pressure, which is, in fact, equivalent to an increase in concentration of the gas.The reaction rate in-

creases in the direction where there are fewer moles of gas and decreases in the reverse direction. For condensed-phase reactions, the pressure dependence is weak.

- Order: The order of the reaction controls how the reactant concentration (or pressure) affects reaction rate.
- Temperature: Usually conducting a reaction at a higher temperature delivers more energy into the system and increases the reaction rate by causing more collisions between particles, as explained by collision theory. However, the main reason that temperature increases the rate of reaction is that more of the colliding particles will have the necessary activation energy resulting in more successful collisions (when bonds are formed between reactants). The influence of temperature is described by the Arrhenius equation.

For example, coal burns in a fireplace in the presence of oxygen, but it does not when it is stored at room temperature. The reaction is spontaneous at low and high temperatures but at room temperature its rate is so slow that it is negligible. The increase in temperature, as created by a match, allows the reaction to start and then it heats itself, because it is exothermic. That is valid for many other fuels, such as methane, butane, and hydrogen.

Reaction rates can be independent of temperature (*non-Arrhenius*) or decrease with increasing temperature (*anti-Arrhenius*). Reactions without an activation barrier (e.g., some radical reactions), tend to have anti Arrhenius temperature dependence: the rate constant decreases with increasing temperature.

- Solvent: Many reactions take place in solution and the properties of the solvent affect the reaction rate. The ionic strength also has an effect on reaction rate.
- Electromagnetic radiation and intensity of light: Electromagnetic radiation is a form of energy. As such, it may speed up the rate or even make a reaction spontaneous as it provides the particles of the reactants with more energy. This energy is in one way or another stored in the reacting particles (it may break bonds, promote molecules to electronically or vibrationally excited states...) creating intermediate species that react easily. As the intensity of light increases, the particles absorb more energy and hence the rate of reaction increases.

For example, when methane reacts with chlorine in the dark, the reaction rate is very slow. It can be sped up when the mixture is put under diffused light. In bright sunlight, the reaction is explosive.

- A catalyst: The presence of a catalyst increases the reaction rate (in both the forward and reverse reactions) by providing an alternative pathway with a lower activation energy.

For example, platinum catalyzes the combustion of hydrogen with oxygen at room temperature.

- Isotopes: The kinetic isotope effect consists in a different reaction rate for the same molecule if it has different isotopes, usually hydrogen isotopes, because of the relative mass difference between hydrogen and deuterium.
- Surface Area: In reactions on surfaces, which take place for example during heterogeneous catalysis, the rate of reaction increases as the surface area does. That is because more particles of the solid are exposed and can be hit by reactant molecules.
- Stirring: Stirring can have a strong effect on the rate of reaction for heterogeneous reactions.
- Diffusion limit: Some reactions are limited by diffusion.

All the factors that affect a reaction rate, except for concentration and reaction order, are taken into account in the reaction rate coefficient (the coefficient in the rate equation of the reaction).

Temperature Dependence

Each reaction rate coefficient k has a temperature dependency, which is usually given by the Arrhenius equation:

$$k = Ae^{-\frac{E_a}{RT}}$$

E_a is the activation energy and R is the gas constant. Since at temperature T the molecules have energies given by a Boltzmann distribution, one can expect the number of collisions with energy greater than E_a to be proportional to $e^{-E_a/R}$. A is the pre-exponential factor or frequency factor.

The values for A and E_a are dependent on the reaction. There are also more complex equations possible, which describe temperature dependence of other rate constants that do not follow this pattern.

A chemical reaction takes place only when the reacting particles collide. However, not all collisions are effective in causing the reaction. Products are formed only when the colliding particles possess a certain minimum energy called threshold energy. As a rule of thumb, reaction rates for many reactions double for every 10 degrees Celsius increase in temperature, For a given reaction, the ratio of its rate constant at a higher temperature to its rate constant at a lower temperature is known as its temperature coefficient (Q). Q_{10} is commonly used as the ratio of rate constants that are 10°C apart.

Pressure Dependence

The pressure dependence of the rate constant for condensed-phase reactions (i.e., when reactants and products are solids or liquid) is usually sufficiently weak in the range of pressures normally encountered in industry that it is neglected in practice.

The pressure dependence of the rate constant is associated with the activation volume. For the reaction proceeding through an activation-state complex:

$$A + B \rightleftharpoons |A \cdots B|^{\ddagger} \rightarrow P$$

the activation volume, $\Delta V^{\ddagger}$, is:

$$\Delta V^{\ddagger} = \bar{V}_{\ddagger} - \bar{V}_{\mathrm{A}} - \bar{V}_{\mathrm{B}}$$

where $\bar{V}$ denotes the partial molar volume of a species and ‡ indicates the activation-state complex.

For the above reaction, one can expect the change of the reaction rate constant (based either on mole fraction or on molar concentration) with pressure at constant temperature to be:

$$\left(\frac{\partial \ln k_x}{\partial P}\right)_T = -\frac{\Delta V^{\ddagger}}{RT}$$

In practice, the matter can be complicated because the partial molar volumes and the activation volume can themselves be a function of pressure.

Reactions can increase or decrease their rates with pressure, depending on the value of $\Delta V^{\ddagger}$. As an example of the possible magnitude of the pressure effect, some organic reactions were shown to double the reaction rate when the pressure was increased from atmospheric (0.1 MPa) to 50 MPa (which gives $\Delta V^{\ddagger} = -0.025$ L/mol).

Rate Equation

The rate of reaction is the change of concentration of a substance in a given time. Whether that be reactants disappearing or products appearing; the rate of reaction is affected by the temperature. However, the chemical equation does not tell us how fast things happen, for this we use a rate equation.

The rate law for a chemical reaction is an equation that relates the reaction rate with the concentrations or partial pressures of the reactants. For the general reaction a $\mathrm{A} + \mathrm{bB} \rightarrow \mathrm{C}$ with no intermediate steps in its reaction mechanism, meaning that it is an elementary reaction, the rate law is given by:

$$r = k[A]^x[B]^y$$

In this equation, [A] and [B] express the concentrations of A and B, respectively, in units of moles per liter. The exponents x and y vary for each reaction, and they must be determined experimentally; they are *not* related to the stoichiometric coefficients of the chemical equation. Lastly, k is known as the rate constant of the reaction. The value of this coefficient k will vary with conditions that affect reaction rate, such as temperature, pressure, surface area, etc. A smaller rate constant indicates a slower reaction, while a larger rate constant indicates a faster reaction.

Reaction	Rate Law
$2NO(g) + O_2(g) \rightarrow 2NO_2(g)$	Rate $= k[NO]^2[O_2]$
$2NO(g) + 2H_2(g) \rightarrow 2N_2(g) + 2H_2O(g)$	Rate $= k[NO]^2[H_2]$
$2ICl(g) + H_2(g) \rightarrow 2HCl(g) + I_2(g)$	Rate $= k[ICl][H_2]$
$2N_2O_5(g) \rightarrow 4NO_2(g) + O_2(g)$	Rate $= k[N_2O_5]$
$2NO_2(g) + F_2(g) \rightarrow 2NO_2F(g)$	Rate $= k[NO_2][F_2]$
$2H_2O_2(aq) \rightarrow 2H_2O(l) + O_2(g)$	Rate $= k[H_2O_2]$
$H_2(g) + Br_2(g) \rightarrow 2HBr(g)$	Rate $= k[H_2][Br_2]^{\frac{1}{2}}$
$O_3(g) + Cl(g) \rightarrow O_2(g) + ClO(g)$	Rate $= k[O_3][Cl]$

Rate laws for various reactions: A variety of reaction orders are observed. Note that the reaction order is unrelated to the stoichiometry of the reactions; it must be determined experimentally.

Reaction Order

To reiterate, the exponents x and y are not derived from the balanced chemical equation, and the rate law of a reaction must be determined experimentally. These exponents may be either integers or fractions, and the *sum* of these exponents is known as the overall reaction order. A reaction can also be described in terms of the order of each reactant. For example, the rate law $Rate = k[NO]^2[O_2]$ describes a reaction which is second-order in nitric oxide, first-order in oxygen, and third-order overall. This is because the value of x is 2, and the value of y is 1, and $2+1=3$.

Example 1

A certain rate law is given as $\text{Rate} = \text{k}[\text{H}_2][\text{Br}_2]^{\frac{1}{2}}$. What is the reaction order?

$$x = 1, y = \frac{1}{2}$$

$$\text{reaction order } = x + y = 1 + \frac{1}{2} = \frac{3}{2}$$

The reaction is first-order in hydrogen, one-half-order in bromine, and $\frac{3}{2}$ order overall.

Example 2

The reaction between nitric oxide and ozone, $\text{NO(g)} + \text{O}_3\text{(g)} \rightarrow \text{NO}_2\text{(g)} + \text{O}_2\text{(g)}$, is first order in both nitric oxide and ozone. The rate law equation for this reaction is: $\text{Rate} = k[NO]^1[O_3]^1$. The overall order of the reaction is $1 + 1 = 2$.

First-order Reactions

A first-order reaction depends on the concentration of one reactant, and the rate law is: $r = -\frac{dA}{dt} = k[A]$.

A first-order reaction depends on the concentration of only one reactant. As such, a first-order reaction is sometimes referred to as a unimolecular reaction. While other reactants can be present, each will be zero-order, since the concentrations of these reactants do not affect the rate. Thus, the rate law for an elementary reaction that is first order with respect to a reactant A is given by:

$$r = -\frac{dA}{dt} = k[A]$$

As usual, *k* is the rate constant, and must have units of concentration/time; in this case it has units of 1/s.

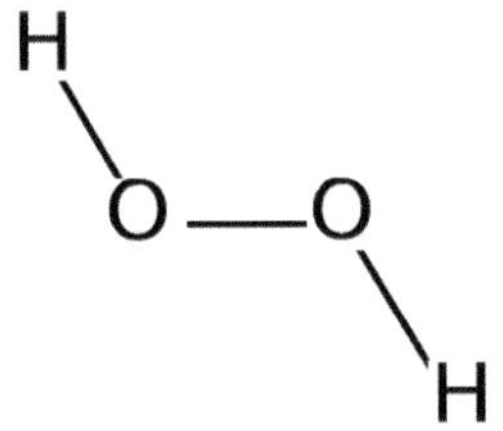

Hydrogen peroxide: The decomposition of hydrogen peroxide to form oxygen and hydrogen is a first-order reaction

Using the Method of Initial Rates to Determine Reaction Order Experimentally

$$2N_2O_5(g) \rightarrow 4NO_2(g) + O_2(g)$$

The balanced chemical equation for the decomposition of dinitrogen pentoxide is given above. Since there is only one reactant, the rate law for this reaction has the general form:

$$Rate = k[N_2O_5]^m$$

In order to determine the overall order of the reaction, we need to determine the value of the exponent *m*. To do this, we can measure an initial concentration of N_2O_5 in a flask, and record the rate at which the N_2O_5 decomposes. We can then run the reaction a second time, but with a different initial concentration of N_2O_5. We then measure the new rate at which the N_2O_5decomposes. By comparing these rates, it is possible for us to find the order of the decomposition reaction.

Example

Let's say that at 25°C, we observe that the rate of decomposition of N_2O_5 is 1.4×10^{-3} M/s when the initial concentration of N_2O_5 is 0.020 M. Then, let's say that we run the experiment again at the same temperature, but this time we begin with a different concentration of N_2O_5, which is 0.010 M. On this second trial, we observe that the rate of decomposition of N_2O_5 is 7.0×10^{-4} M/s. We can now set up a ratio of the first rate to the second rate:

$$\frac{Rate_1}{Rate_2} = \frac{k[N_2O_5]_{i1}^m}{k[N_2O_5]_{i2}^m}$$

$$\frac{1.4\times10^{-3}}{7.0\times10^{-4}} = \frac{k(0.020)^m}{k(0.010)_m}$$

Notice that the left side of the equation is simply equal to 2, and that the rate constants cancel on the right side of the equation. Everything simplifies to:

$$2.0 = 2.0^m$$

Clearly, then, m=1, and the decomposition is a first-order reaction.

Determining the Rate Constant *k*

Once we have determined the order of the reaction, we can go back and plug in one set of our initial values and solve for *k*. We find that:

$$rate = k[N_2O_5]^1 = k[N_2O_5]$$

Substituting in our first set of values, we have,

$$1.4\times10^{-3} = k(0.020)$$

$$k = 0.070s^{-1}$$

Second-order Reactions

A second-order reaction is second-order in only one reactant, or first-order in two reactants.

A reaction is said to be second-order when the overall order is two. For a reaction with the general form aA+bB→CaA+bB→C, the reaction can be second order in two possible ways. It can be second-order in either A or B, or first-order in both A and B. If the reaction were second-order in either reactant, it would lead to the following rate laws:

$$rate = k[A]^2$$

or,

$$rate = k[B]^2$$

The second scenario, in which the reaction is first-order in both A and B, would yield the following rate law:

$$rate = k[A][B]$$

If we are interested in determining the order of the reaction with respect to A and B, we apply the method of initial rates.

Determining Reaction Order in A

In order to determine the reaction order for A, we can set up our first equation as follows:

$$\frac{r_1}{r_2} = \frac{k[A]_1^x[B]_1^y}{k[A]_2^x[B]_2^y}$$

$$\frac{5.46}{12.28} = \frac{k(0.200)^x(0.200)^y}{k(0.300)^x(0.200)^y}$$

Note that on the right side of the equation, both the rate constant *k* and the term (0.200)y(0.200)y cancel. This was done intentionally, because in order to determine the reaction order in A, we need to choose two experimental trials in which the initial concentration of A changes, but the initial concentration of B is constant, so that the concentration of B cancels. Our equation simplifies to:

$$\frac{5.46}{12.28} = \frac{(0.200)^x}{(0.300)^x}$$

$$0.444 = \left(\frac{2}{3}\right)^x$$

$$\ln(0.444) = x \cdot \ln\left(\frac{2}{3}\right)$$

$$x \approx 2$$

Therefore, the reaction is second-order in A.

Determining Reaction Order in B

Next, we need to determine the reaction order for B. We do this by picking two trials in which the concentration of B changes, but the concentration of A does not. Trials 1 and 3 will do this for us, and we set up our ratios as follows:

$$\frac{x_1}{x_3} = \frac{k[A]_1^2 [B]_1^y}{k[A]_3^2 [B]_3^y}$$

$$\frac{5.46}{5.42} = \frac{k(0.200)^2(0.200)^y}{k(0.200)^2(0.400)^y}$$

Note that both k and the concentrations of A cancel. Also, $\frac{5.46}{5.42} \approx 2$, so everything simplifies to:

$$1 = \frac{(0.200)^y}{(0.400)^y}$$

$$1 = \left(\frac{1}{2}\right)^y$$

y=0

Therefore, the reaction is zero-order in B.

Overall Reaction Order

We have determined that the reaction is second-order in A, and zero-order in B. Therefore, the overall order for the reaction is second-order $(2+0=2)$, and the rate law will be:

$$rate = k[A]^2$$

Zero-order Reactions

A zero-order reaction has a constant rate that is independent of the concentration of the reactant(s); the rate law is simply $\text{rate} = \text{k}$.

Unlike the other orders of reaction, a zero-order reaction has a rate that is independent of the concentration of the reactant(s). As such, increasing or decreasing the concentration of the reacting species will not speed up or slow down the reaction rate. Zero-order reactions are typically found when a material that is required for the reaction to proceed, such as a surface or a catalyst, is saturated by the reactants.

The rate law for a zero-order reaction is $rate = k$, where k is the rate constant. In the case of a zero-order reaction, the rate constant *k*will have units of concentration/time, such as *M/s*.

Plot of Concentration versus Time for a Zero-order Reaction

Recall that the rate of a chemical reaction is defined in terms of the change in concentration of a reactant per change in time. This can be expressed as follows:

$$rate = -\frac{d[A]}{dt} = k$$

By rearranging this equation and using a bit of calculus, we get the equation:

$$[A] = -kt$$

This is the integrated rate law for a zero-order reaction. Note that this equation has the form y=m. Therefore, a plot of [A] versus *t*will always yield a straight line with a slope of −k.

Half-life of a Zero-order Reaction

The half-life of a reaction describes the time needed for half of the reactant(s) to be depleted, which is the same as the half-life involved in nuclear decay, a first-order reaction. For a zero-order reaction, the half-life is given by:

$$t_{\frac{1}{2}} = \frac{[A]_0}{2k}$$

$[A]_0$ represents the initial concentration and k is the zero-order rate constant.

Example of a Zero-order Reaction

The Haber process is a well-known process used to manufacture ammonia from hydrogen and nitrogen gas. The reverse of this is known, simply, as the reverse Haber process, and it is given by:

$$2NH_3(g) \rightarrow 3H_2(g) + N_2(g)$$

The reverse Haber process is an example of a zero-order reaction because its rate is independent of the concentration of ammonia. As always, it should be noted that the order of this reaction, like the order for all chemical reactions, cannot be deduced from the chemical equation, but must be determined experimentally.

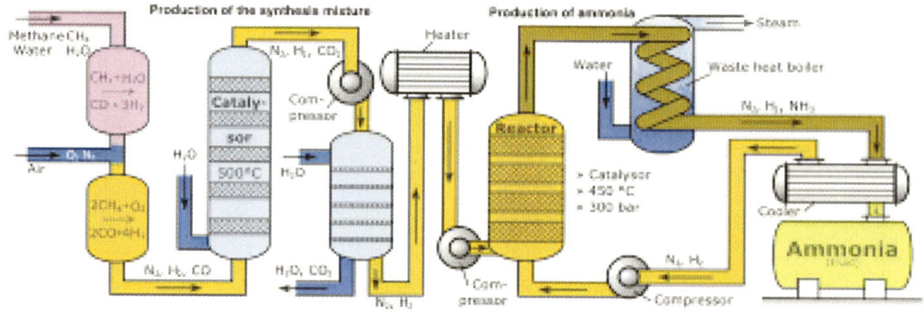

The Haber process produces ammonia from hydrogen and nitrogen gas. The reverse of this process (the decomposition of ammonia to form nitrogen and hydrogen) is a zero-order reaction

The Integrated Rate Law

The integrated rate laws derive from calculus, and they relate the concentrations of reactants with time.

The rate law is a differential equation, meaning that it describes the change in concentration of reactant (s) per change in time. Using calculus, the rate law can be integrated to obtain an integrated rate equation that links concentrations of reactants or products with time directly.

Integrated Raw Law for a First-order Reaction

Recall that the rate law for a first-order reaction is given by:

$$rate = \frac{-d[A]}{dt} = k[A]$$

We can rearrange this equation to combine our variables, and integrate both sides to get our integrated rate law:

$$\int_{[A]_0}^{[A]_t} \frac{d[A]}{[A]} = -\int_0^t k\,dt$$

$$\ln\left(\frac{[A]_t}{[A]_0}\right) = -kt$$

$$\frac{[A]_t}{[A]_0} = e^{-kt}$$

Finally, putting this equation in terms of $[A]_t$, we have:

$$[A]_t = [A]_0 e^{-kt}$$

This is the final form of the integrated rate law for a first-order reaction. Here, $[A]_t$ represents the concentration of the chemical of interest at a particular time *t*, and $[A]_0$ represents the initial concentration of A. Note that this equation can also be written in the following form:

$$\ln[A] = -kt + ln[A]_0$$

This form is useful, because it is of the form y=mx+b. When the integrated rate law is written in this way, a plot of ln[A] versus *t* will yield a straight line with the slope -*k*. However, the integrated first-order rate law is usually written in the form of the exponential decay equation.

Integrated Rate Law for a Second-order Reaction

Recall that the rate law for a second-order reaction is given by:

$$rate = -\frac{d[A]}{dt} = k[A]^2$$

Rearranging our variables and integrating, we get the following:

$$\int_{[A]_0}^{[A]_t} \frac{d[A]}{[A]^2} = -\int_0^t k \ dt$$

$$\frac{1}{[A]_t} - \frac{1}{[A]_0} = kt$$

$\frac{1}{[A]_t} = \frac{1}{[A]_0} + kt$ The final version of this integrated rate law is given by:

Note that this equation is also of the form y=mx+b. Here, a plot of $\frac{1}{[A]_t}$ versus *t* will yield a straight line with a positive slope *k*.

Integrated Rate Law for Second-order Reaction with Two Reactants

For a reaction that is second-order overall, and first-order in two reactants, A and B, our rate law is given by:

$$rate = -\frac{d[A]}{dt} = -\frac{d[B]}{dt} = k[A][B]$$

There are two possible scenarios here. The first is that the initial concentrations of A and B are equal, which simplifies things greatly. In this case, we can say that [A]=[B], and the rate law simplifies to:

$$rate = k[A]^2$$

This is the standard form for second-order rate law, and the integrated rate law will be the same as above. However, in the case where $[A]_0 \neq [B]_0$, the integrated rate law will take the form:

$$ln\frac{[B][A]_0}{[A][B]_0} = k([B]_0 - [A]_0)t$$

In this more complicated instance, a plot of $\ln\frac{[B][A]_0}{[A][B]_0}$ versus *t* will yield a straight line with a slope of $k([B]_0 - [A]_0)$.

Integrated Rate Law for a Zero-order Reaction

The rate law for a zero-order reaction is given by:

$$rate = -\frac{d[A]}{dt} = k$$

Rearranging and integrating, we have:

$$\int_{[A]_0}^{[A]_t} d[A] = -\int_0^t k\, dt$$

$$[A]_t - [A]_0 = -kt$$

$$[A]_t = -kt + [A]_0$$

Note here that a plot of [A] versus *t* will yield a straight line with the slope -*k*. The y-intercept of this plot will be the initial concentration of A, $[A]_0$.

Half-life

Number of half-lives elapsed	Fraction remaining	Percentage remaining
0	$^1/_1$	100
1	$^1/_2$	50
2	$^1/_4$	25
3	$^1/_8$	12.5
4	$^1/_{16}$	6.25
5	$^1/_{32}$	3.125
6	$^1/_{64}$	1.563
7	$^1/_{128}$	0.781
...	...	...
n	2^{-n}	$100/(2^n)$

The half-life of a reaction is the amount of time it takes for it to become half its quantity.

The half-life of a reaction is the amount of time it takes for the concentration of a reactant to decrease to one-half of its initial value.

The half-life is the time required for a quantity to fall to half its initial value, as measured at the beginning of the time period. If we know the integrated rate laws, we can determine the half-lives for first-, second-, and zero-order reactions. For this discussion, we will focus on reactions with a single reactant.

Half-Life of a First-order Reaction

Recall that for a first-order reaction, the integrated rate law is given by:

$$[A] = [A]_0 e^{-(kt)}$$

This can be written another way, equivalently:

$$\ln[A] = ln[A]_0 - kt$$

If we are interested in finding the half-life for this reaction, then we need to solve for the time at which the concentration, [A], is equal to half of what it was initially; that is, $\frac{[A]_0}{2}$. If we plug this in for [A] in our integrated rate law, we have:

$$ln\frac{[A]_0}{2} = ln[A]_0 - kt$$

By rearranging this equation and using the properties of logarithms, we can find that, for a first order reaction:

$$t_{\frac{1}{2}} = \frac{ln(2)}{k}$$

What is interesting about this equation is that it tells us that the half-life of a first-order reaction does not depend on how much material we have at the start. It takes exactly the same amount of time for the reaction to proceed from all of the starting material to half of the starting material as it does to proceed from half of the starting material to one-fourth of the starting material. In each case, we halve the remaining material in a time equal to the constant half-life. Keep in mind that these conclusions are only valid for first-order reactions.

Consider, for example, a first-order reaction that has a rate constant of 5.00 s^{-1}. To find the half-life of the reaction, we would simply plug 5.00 s^{-1} in for *k*:

$$t_{\frac{1}{2}} = \frac{ln(2)}{k}$$

$$t_{\frac{1}{2}} = \frac{ln(2)}{5.00s^{-1}} = 0.14s$$

Half-life for Second-order Reactions

Recall our integrated rate law for a second-order reaction:

$$\frac{1}{[A]} = \frac{1}{[A]_0} + kt$$

To find the half-life, we once again plug in $\frac{[A]_0}{2}$ for [A].

$$\frac{1}{\frac{[A]_0}{2}} = \frac{1}{[A]_0} + kt$$

Solving for *t*, we get:

$$t_{\frac{1}{2}} = \frac{1}{K[A]_0}$$

Thus the half-life of a second-order reaction, unlike the half-life for a first-order reaction, does depend upon the initial concentration of A. Specifically, there is an inversely proportional relationship between $t_{\frac{1}{2}}$ and $[A]_0$; as the initial concentration of A increases, the half-life decreases.

Consider, for example, a second-order reaction with a rate constant of 3 M^{-1} s^{-1} in which the initial concentration of A is 0.5 M:

$$t_{\frac{1}{2}} = \frac{1}{K[A]_0} \quad t_{\frac{1}{2}} = \frac{1}{(3)(0.5)} = 0.67\ s$$

Half-life for a Zero-order Reaction

The integrated rate law for a zero-order reaction is given by:

$$[A] = [A]_0 - k$$

Subbing in $\frac{[A]_0}{2}$ for $[A]$, we have:

$$\frac{[A]_0}{2} = [A]_0 - k$$

Rearranging in terms of t, we can obtain an expression for the half-life:

$$t_{\frac{1}{2}} = \frac{[A]_0}{2k}$$

Therefore, for a zero-order reaction, half-life and initial concentration are directly proportional. As initial concentration increases, the half-life for the reaction gets longer and longer.

Fractional Order

In fractional order reactions, the order is a non-integer, which often indicates a chemical chain reaction or other complex reaction mechanism. For example, the pyrolysis of acetaldehyde (CH_3CHO) into methane and carbon monoxide proceeds with an order of 1.5 with respect to acetaldehyde: $r=k[CH_3CHO]^{3/2}$. The decomposition of phosgene ($COCl_2$) to carbon monoxide and chlorine has order 1 with respect to phosgene itself and order 0.5 with respect to chlorine: $r = k[COCl_2]\,[Cl_2]^{1/2}$.

The order of a chain reaction can be rationalized using the steady state approximation for the concentration of reactive intermediates such as free radicals. For the pyrolysis of acetaldehyde, the Rice-Herzfeld mechanism is:

Initiation

$$CH_3CHO \rightarrow \bullet CH_3 + \bullet CHO$$

Propagation

$$\bullet CH_3 + CH_3CHO \rightarrow CH_3CO\bullet + CH_4$$

$$CH_3CO\bullet \rightarrow \bullet CH_3 + CO$$

Termination

$$2\ \bullet CH_3 \rightarrow C_2H_6$$

where • denotes a free radical. To simplify the theory, the reactions of the •CHO to form a second $\bullet CH_3$ are ignored.

In the steady state, the rates of formation and destruction of methyl radicals are equal, so that

$$\frac{d[.CH3]}{dt} = k_i[CH3CHO] - k_t[.CH3]^2 = 0,$$

so that the concentration of methyl radical satisfies

$$[.CH3] \propto [CH3CHO]^{1/2}.$$

The reaction rate equals the rate of the propagation steps which form the main reaction products CH_4 and CO:

$$v = \frac{d[CH4]}{dt} = k_p[.CH3][CH3CHO] \propto [CH3CHO]^{3/2}$$

in agreement with the experimental order of 3/2.

Mixed Order

More complex rate laws have been described as being *mixed order* if they approximate to the laws for more than one order at different concentrations of the chemical species involved. For example, a rate law of the form represents concurrent first order and second order reactions (or more often concurrent pseudo-first order and second order) reactions, and can be described as mixed first and second order. For sufficiently large values of [A] such a reaction will approximate second order kinetics, but for smaller [A] the kinetics will approximate first order (or pseudo-first order). As the reaction progresses, the reaction can change from second order to first order as reactant is consumed.

Another type of mixed-order rate law has a denominator of two or more terms, often because the identity of the rate-determining step depends on the values of the concentrations. An example is the oxidation of an alcohol to a ketoneby hexacyanoferrate (III) ion [$Fe(CN)_6^{3-}$] with ruthenate (VI) ion (RuO_4^{2-}) as catalyst. For this reaction, the rate of disappearance of hexacyanoferrate (III) is $r = \frac{[Fe(CN)6]^2 -}{k_\alpha + k_\beta[Fe(CN)6]^2 -}$.

This is zero-order with respect to hexacyanoferrate (III) at the onset of the reaction (when its concentration is high and the ruthenium catalyst is quickly regenerated), but changes to first-order when its concentration decreases and the regeneration of catalyst becomes rate-determining.

Notable mechanisms with mixed-order rate laws with two-term denominators include:

- Michaelis-Menten kinetics for enzyme-catalysis: first-order in substrate (second-order overall) at low substrate concentrations, zero order in substrate (first-order overall) at higher substrate concentrations; and
- The Lindemann mechanism for unimolecular reactions: second-order at low pressures, first-order at high pressures.

Negative Order

A reaction rate can have a negative partial order with respect to a substance. For example, the conversion of ozone (O_3) to oxygen follows the rate equation $r = k\frac{[O_3]^2}{[O_2]}$ in an excess of oxygen. This corresponds to second order in ozone and order (-1) with respect to oxygen.

When a partial order is negative, the overall order is usually considered as undefined. In the above example for instance, the reaction is not described as first order even though the sum of the partial orders is 2 + (−1) = 1, because the rate equation is more complex than that of a simple first-order reaction.

Opposed Reactions

A pair of forward and reverse reactions may occur simultaneously with comparable speeds. For example, A and B react into X and Y and vice versa (*s, t, u*, and *v* are the stoichiometric coefficients):

$$A + tB \rightleftharpoons uX + vY$$

The reaction rate expression for the above reactions (assuming each one is elementary) can be expressed as:

$$r = k_1[A]^s[B]^t - k_2[X]^u[Y]^v$$

where: k_1 is the rate coefficient for the reaction that consumes A and B; k_2 is the rate coefficient for the backwards reaction, which consumes X and Y and produces A and B.

The constants k_1 and k_2 are related to the equilibrium coefficient for the reaction (K) by the following relationship (set *r*=0 in balance):

$$k_1[A]^s[B]^t = k_2[X]^u[Y]^v$$

$$K = \frac{[X]^u[Y]^v}{[A]^s[B]^t} = \frac{k_1}{k_2}$$

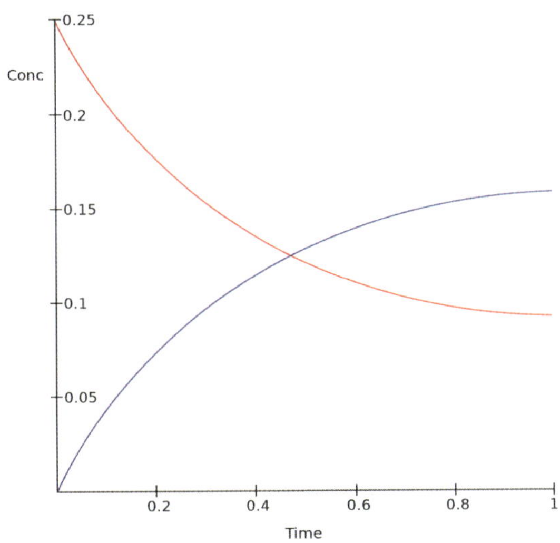

Concentration of A (A_0 = 0.25 mole/l) and B versus time reaching equilibrium $k_f = 2\ min^{-1}$ and $k_r = 1\ min^{-1}$

Simple Example

In a simple equilibrium between two species:

$$A \rightleftharpoons B$$

Where the reactions starts with an initial concentration of A, $[A]_0$, with an initial concentration of 0 for B at time t=0.

Then the constant K at equilibrium is expressed as:

$$K \overset{\text{def}}{=} \frac{k_f}{k_b} = \frac{[B]_e}{[A]_e}$$

Where $[A]_e$ and $[B]_e$ are the concentrations of A and B at equilibrium, respectively.

The concentration of A at time t, $[A]_t$, is related to the concentration of B at time t, $[B]_t$, by the equilibrium reaction equation:

$$[A]_t = [A]0 - [B]_t$$

Note that the term $[B]_0$ is not present because, in this simple example, the initial concentration of B is 0.

This applies even when time t is at infinity; i.e., equilibrium has been reached:

$$[A]_e = [A]_0 - [B]_e$$

then it follows, by the definition of K, that,

$$[B]_e = x = \frac{k_f}{k_f + k_b}[A]_0$$

and, therefore,

$$[A]_e = [A]_0 - x = \frac{k_b}{k_f + k_b}[A]_0$$

These equations allow us to uncouple the system of differential equations, and allow us to solve for the concentration of A alone.

The reaction equation, given previously as:

$$r = k_1[A]^s[B]^t - k_2[X]^u[Y]^v$$

$$-\frac{d[A]}{dt} = k_f[A]_t - k_b[B]_t$$

The derivative is negative because this is the rate of the reaction going from A to B, and therefore the concentration of A is decreasing. To simplify annotation, let x be $[A]_t$, the concentration of A at time t. Let x_e be the concentration of A at equilibrium. Then:

$$-\frac{d[A]}{dt}=k_f[A]_t-k_b[B]_t$$

$$-\frac{dx}{dt}=k_f x-k_b[B]_t$$

$$=k_f x-k_b([A]_0-x)$$

$$=(k_f+k_b)x-k_b[A]_0$$

Since:

$$k_f+k_b=k_b\frac{[A]_0}{x_e}$$

The reaction rate becomes:

$$\frac{dx}{dt}=\frac{k_b[A]_0}{x_e}(x_e-x)$$

which results in:

$$\ln\left(\frac{[A]_0-[A]_e}{[A]_t-[A]_e}\right)=(k_f+k_b)t$$

A plot of the negative natural logarithm of the concentration of A in time minus the concentration at equilibrium versus time t gives a straight line with slope $k_f + k_b$. By measurement of A_e and B_e the values of K and the two reaction rate constants will be known.

Generalization of Simple Example

If the concentration at the time $t = 0$ is different from above, the simplifications above are invalid, and a system of differential equations must be solved. However, this system can also be solved exactly to yield the following generalized expressions:

$$[A]=[A]_0\frac{1}{k_f+k_b}\left(k_b+k_f e^{-(k_f+k_b)t}\right)+[B]_0\frac{k_b}{k_f+k_b}\left(1-e^{-(k_f+k_b)t}\right)$$

$$[B]=[A]_0\frac{k_f}{k_f+k_b}\left(1-e^{-(k_f+k_b)t}\right)+[B]_0\frac{1}{k_f+k_b}\left(k_f+k_b e^{-(k_f+k_b)t}\right)$$

When the equilibrium constant is close to unity and the reaction rates very fast for instance in conformational analysis of molecules, other methods are required for the determination of rate constants for instance by complete lineshape analysis in NMR spectroscopy.

Consecutive Reactions

If the rate constants for the following reaction are k_1 and k_2; $A \rightarrow B \rightarrow C$, then the rate equation is:

For reactant A: $\frac{d[A]}{dt} = -k_1[A]$

For reactant B: $\frac{d[B]}{dt} = k_1[A] - k_2[B]$

For product C: $\frac{d[C]}{dt} = k_2[B]$

With the individual concentrations scaled by the total population of reactants to become probabilities, linear systems of differential equations such as these can be formulated as a master equation. The differential equations can be solved analytically and the integrated rate equations are:

$$[A] = [A]_0 e^{-k_1 t}$$

$$[B] = \begin{cases} [A]_0 \frac{k_1}{k_2 - k_1}\left(e^{-k_1 t} - e^{-k_2 t}\right) + [B]_0 e^{-k_2 t} & k_1 \neq k_2 \\ [A]_0 k_1 t e^{-k_1 t} + [B]_0 e^{-k_1 t} & \text{otherwise,} \end{cases}$$

$$[C] = \begin{cases} [A]_0 \left(1 + \frac{k_1 e^{-k_2 t} - k_2 e^{-k_1 t}}{k_2 - k_1}\right) + [B]_0 \left(1 - e^{-k_2 t}\right) + [C]_0 & k_1 \neq k_2 \\ [A]_0 \left(1 - e^{-k_1 t} - k_1 t e^{-k_1 t}\right) + [B]_0 \left(1 - e^{-k_1 t}\right) + [C]_0 \end{cases}$$

The steady state approximation leads to very similar results in an easier way.

Parallel or Competitive Reactions

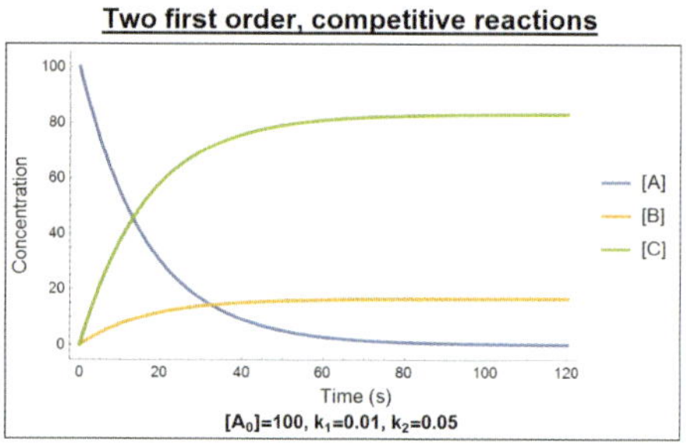

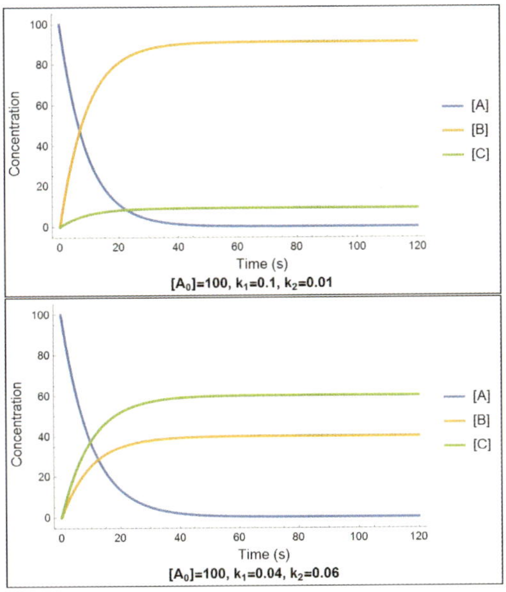

Time course of two first order, competitive reactions with differing rate constants

When a substance reacts simultaneously to give two different products, a parallel or competitive reaction is said to take place.

Two First Order Reactions

$A \to B$ and $A \to C$, with constants k_1 and k_2 and rate equations $-\dfrac{d[A]}{dt} = (k_1 + k_2)[A]$; $\dfrac{d[B]}{dt} = k_1[A]$ and $\dfrac{d[C]}{dt} = k_2[A]$.

The integrated rate equations are then $[A] = [A]_0 e^{-(k_1+k_2)t}$; $[B] = \dfrac{k_1}{k_1 + k_2}[A]_0(1 - e^{-(k_1+k_2)t})$ and $[C] = \dfrac{k_2}{k_1 + k_2}[A]_0(1 - e^{-(k_1+k_2)t})$.

One important relationship in this case is $\dfrac{[B]}{[C]} = \dfrac{k_1}{k_2}$

One First Order and One Second Order Reaction

This can be the case when studying a bimolecular reaction and a simultaneous hydrolysis (which can be treated as pseudo order one) takes place: the hydrolysis complicates the study of the reaction kinetics, because some reactant is being "spent" in a parallel reaction. For example, A reacts with R to give our product C, but meanwhile the hydrolysis reaction takes away an amount of A to give B, a byproduct: $A + H_2O \to B$ and $A + R \to C$. The rate equations are: $\dfrac{d[B]}{dt} = k_1[A][H_2O] = k_{1'}[A]$ and $\dfrac{d[C]}{dt} = k_2[A][R]$ Where $k_{1'}$ is the pseudo first order constant.

The integrated rate equation for the main product [C] is $[C]=[R]_0\left[1-e^{-\frac{k_2}{k_{1'}}[A]_0(1-e^{-k_{1'}t})}\right]$,

which is equivalent to $\ln\frac{[R]_0}{[R]_0-[C]}=\frac{k_2[A]_0}{k_{1'}}(1-e^{-k_{1'}t})$. Concentration of B is related to that of C through $[B]=-\frac{k_{1'}}{k_2}ln\left(1-\frac{[C]}{[R]_0}\right)$.

The integrated equations were analytically obtained but during the process it was assumed that $[A]_0-[C]\approx[A]_0$ therefeore, previous equation for $[C]$ can only be used for low concentrations of [C] compared to $[A]_0$.

Stoichiometric Reaction Networks

The most general description of a chemical reaction network considers a number N of distinct chemical species reacting via R reactions. The chemical equation of the j-th reaction can then be written in the generic form,

$$s_{1j}X_1+s_{2j}X_2\ldots+s_{Nj}X_N\xrightarrow{k_j}r_{1j}X_1+r_{2j}X_2+\ldots+r_{Nj}X_N,$$

which is often written in the equivalent form,

$$\sum_{i=1}^{N}s_{ij}X_i\xrightarrow{k_j}\sum_{i=1}^{N}r_{ij}X_i.$$

Here

j is the reaction index running from 1 to R,

X_i denotes the i-th chemical species,

k_j is the rate constant of the j-th reaction and

s_{ij} and r_{ij} are the stoichiometric coefficients of reactants and products, respectively.

The rate of such reaction can be inferred by the law of mass action,

$$f_j\left(\left[\vec{X}\right]\right)=K_j\prod_{z=1}^{N}[X_z]^{s_{zj}}$$

which denotes the flux of molecules per unit time and unit volume. Here $[\vec{X}]=([X1],[X2],\ldots,[X_N])$ is the vector of concentrations. Note that this definition includes the elementary reactions:

- Zero order reactions

for which $s_{zj} = 0$ for all z,

- First order reactions

for which $s_{zj} = 1$ for a single z,

- Second order reactions

for which $s_{zj} = 1$ for exactly two z, i.e., a bimolecular reaction, or $s_{zj} = 2$ for a single z, i.e., a dimerization reaction.

One can define the stoichiometric matrix,

$$S_{ij} = r_{ij} - s_{ij},$$

denoting the net extent of molecules of i in reaction j. The reaction rate equations can then be written in the general form,

$$\frac{d[X_i]}{dt} = \sum_{j=1}^{R} S_{ij} f_j([\vec{X}]).$$

Note that this is the product of the stoichiometric matrix and the vector of reaction rate functions. Particular simple solutions exist in equilibrium, $\frac{d[X_i]}{dt} = 0$, for systems composed of merely reversible reactions. In this case the rate of the forward and backward reactions are equal, a principle called detailed balance. Note that detailed balance is a property of the stoichiometric matrix S_{ij} alone and does not depend on the particular form of the rate functions f_j. All other cases where detailed balance is violated are commonly studied by flux balance analysis which has been developed to understand metabolic pathways.

General Dynamics of Unimolecular Conversion

For a general unimolecular reaction involving interconversion of N different species, whose concentrations at time t are denoted by $X_1(t)$ through $X_N(t)$, an analytic form for the time-evolution of the species can be found. Let the rate constant of conversion from species X_i to species X_j be denoted as k_{ij},, and construct a rate-constant matrix K whose entries are the k_{ij}.

Also, let $X(t) = (X_1(t), X_2(t), ..., X_N(t))^T$ be the vector of concentrations as a function of time.

Let $J = (1,1,1,...,1)^T$ be the vector of ones.

Let I be the $N \times N$ identity matrix.

Let $Diag$ be the function that takes a vector and constructs a diagonal matrix whose on-diagonal entries are those of the vector.

Let $\mathcal{L}^{-1}$ be the inverse Laplace transform from s to t.

Then the time-evolved state $X(t)$ is given by,

$$X(t)=\mathcal{L}^{-1}[(sI+Diag(KJ)-K^{T})^{-1}X(0)],$$

thus providing the relation between the initial conditions of the system and its state at time t.

Rate-determining Step

The rate determining step is the slowest step of a chemical reaction that determines the speed (rate) at which the overall reaction proceeds. The rate determining step can be compared to the neck of a funnel. The rate at which water flows through a funnel is limited / determined by the width of the neck of the funnel and not by the rate at which the water is poured into the funnel. Like the neck of the funnel, the slow step of a reaction determines the rate of a reaction. Not all reactions have rate determining steps and has one only if one step is significantly slower than the other steps in the reaction.

The rate determining step is important in deriving the rate equation of a chemical reaction. For example, consider a multi-step reaction:

$$A+B\rightarrow C+D$$

Assume the elementary steps for this reaction are the following:

Step 1: Slow $A+A\rightarrow C+E$ (with a rate constant, K_1)

Step 2: Fast $E+B\rightarrow A+D$ (with a rate constant, k_2)

where E is an intermediate, the product in step 1 and a reactant in step 2 that does not show up in the overall reaction. This is because when steps 1 and 2 are added, intermediate E cancels out, along with the extra reactant A from step 1. Note that intermediate reactions do not show up in the overall reaction.

In the case of this hypothetical reaction, if step 1 is the slow step and step 2 is the fast step of the reaction, then the overall reaction rate depends on step 1; the slow step in a reaction is always the rate-limiting step, another name for the rate determining step.

Therefore the rate equation is,

$$rate=k_1[A][A]=k1[A]^2$$

Example Reaction: NO2 + CO

As an example, consider the gas-phase reaction $NO_2 + CO \rightarrow NO + CO_2$. If this reaction occurred in a single step, its reaction rate (r) would be proportional to the rate of collisions between NO_2 and CO molecules: $r = k[NO_2][CO]$, where k is the reaction rate constant, and square brackets indicate a molar concentration. Another typical example is the Zel'dovich mechanism.

First Step Rate-determining

In fact, however, the observed reaction rate is second-order in NO_2 and zero-order in CO, with rate equation $r = k[NO_2]^2$. This suggests that the rate is determined by a step in which two NO_2 molecules react, with the CO molecule entering at another, faster, step. A possible mechanism in two elementary steps that explains the rate equation is:

1. $NO_2 + NO_2 \rightarrow NO + NO_3$ *(slow step, rate-determining)*

2. $NO_3 + CO \rightarrow NO_2 + CO_2$ *(fast step)*

In this mechanism the reactive intermediate species NO_3 is formed in the first step with rate r_1 and reacts with CO in the second step with rate r_2. However NO_3 can also react with NO if the first step occurs in the *reverse direction* $(NO + NO_3 \rightarrow 2\,NO_2)$ with rate r_{-1}, where the minus sign indicates the rate of a reverse reaction.

The concentration of a reactive intermediate such as $[NO_3]$ remains low and almost constant. It may therefore be estimated by the steady-state approximation, which specifies that the rate at which it is formed equals the (total) rate at which it is consumed. In this example NO_3 is formed in one step and reacts in two, so that,

$$\frac{d[NO3]}{dt} = r_1 - r_2 - r_{-1} \approx 0.$$

The statement that the first step is the slow step actually means that the first step *in the reverse direction* is slower than the second step in the forward direction, so that almost all NO_3 is consumed by reaction with CO and not with NO. That is, $r_{-1} \ll r_2$, so that $r_1 - r_2 \approx 0$. But the overall rate of reaction is the rate of formation of final product (here CO_2), so that $r = r_2 \approx r_1$.. That is, the overall rate is determined by the rate of the first step, and (almost) all molecules that react at the first step continue to the fast second step.

Pre-equilibrium: if the Second Step were Rate-determining

The other possible case would be that the second step is slow and rate-determining, meaning that it is slower than the first step in the reverse direction: $r_2 \ll r_{-1}$. In this hypothesis, $r_1 - r_{-1} \approx 0$, so that the first step is (almost) at equilibrium. The overall rate is determined by the second step: $r = r_2 \ll r_1$, as very few molecules that react at the first step continue to the second step, which is much slower. Such a situation in which an intermediate (here NO_3) forms an equilibrium with reactants *prior* to the rate-determining step is described as a *pre-equilibrium* For the reaction of NO_2 and

CO, this hypothesis can be rejected, since it implies a rate equation that disagrees with experiment.

If the first step were at equilibrium, then its equilibrium constant expression permits calculation of the concentration of the intermediate NO_3 in terms of more stable (and more easily measured) reactant and product species:

$$K_1 = \frac{[NO][NO_3]}{[NO_2]^2},$$

$$[NO_3] = K_1 \frac{[NO_2]^2}{[NO]}.$$

The overall reaction rate would then be

$$r = r_2 = k_2[NO_3][CO] = k_2K_1 \frac{[NO_2]^2[CO]}{[NO]},$$

which disagrees with the experimental rate law given above, and so disproves the hypothesis that the second step is rate-determining for this reaction. However, some other reactions are believed to involve rapid pre-equilibria prior to the rate-determining step.

Nucleophilic Substitution

Another example is the unimolecular nucleophilic substitution (S_N1) reaction in organic chemistry, where it is the first, rate-determining step that is unimolecular. A specific case is the basic hydrolysis of tert-butyl bromide ($t-C_4H_9Br$) by aqueous sodium hydroxide. The mechanism has two steps (where R denotes the tert-butyl radical $t-C_4H_9$):

1. Formation of a carbocation $R-Br \rightarrow R^+ + Br^-$.
2. Nucleophilic attack by one water molecule $R^+ + OH^- \rightarrow ROH$.

This reaction is found to be first-order with $r=k[R-Br]$, which indicates that the first step is slow and determines the rate. The second step with OH^- is much faster, so the overall rate is independent of the concentration of OH^-.

In contrast, the alkaline hydrolysis of methyl bromide (CH_3Br) is a bimolecular nucleophilic substitution (S_N2) reaction in a single bimolecular step. Its rate law is second-order: $r=k[R-Br][OH^-]$.

Composition of the Transition State

A useful rule in the determination of mechanism is that the concentration factors in the rate law indicate the composition and charge of the activated complex or transition state. For the NO_2–CO reaction above, the rate depends on $[NO_2]^2$, so that the activated complex has composition N_2O_4, with $2NO_2$ entering the reaction before the transition state, and CO reacting after the transition state.

A multistep example is the reaction between oxalic acid and chlorine in aqueous solution: $H_2C_2O_4 + Cl_2 \rightarrow 2CO_2 + 2H^+ + 2Cl^-$. The observed rate law is,

$$v = k\frac{[Cl_2][H2C_2O_4]}{[H+]^2[Cl^-]},$$

which implies an activated complex in which the reactants lose $2H^+ + Cl$ before the rate-determining step. The formula of the activated complex is $Cl_2 + H_2C_2O_4 - 2H^+ - Cl^- + x\,H_2O,\ or\ C_2O_4Cl(H_2O)^-_x$ (an unknown number of water molecules are added because the possible dependence of the reaction rate on H_2O was not studied, since the data were obtained in water solvent at a large and essentially unvarying concentration).

One possible mechanism in which the preliminary steps are assumed to be rapid pre-equilibria occurring prior to the transition state is,

$$Cl_2 + H_2O \rightleftharpoons HOCl + Cl^- + H^+$$
$$H_2C_2O_4 \rightleftharpoons H^+ + HC_2O^-{}_4$$
$$HOCl + HC_2O_4^- \rightarrow H_2O + Cl^- + 2CO_2$$

Reaction Coordinate Diagram

In a multistep reaction, the rate-determining step does not necessarily correspond to the highest Gibbs energy on the reaction coordinate diagram. If there is a reaction intermediate whose energy is lower than the initial reactants, then the activation energy needed to pass through any subsequent transition state depends on the Gibbs energy of that state relative to the lower-energy intermediate. The rate-determining step is then the step with the largest Gibbs energy difference relative either to the starting material or to any previous intermediate on the diagram.

Also, for reaction steps that are not first-order, concentration terms must be considered in choosing the rate-determining step.

Chain Reactions

Not all reactions have a single rate-determining step. In particular, the rate of a chain reaction is usually not controlled by any single step.

Diffusion Control

In the previous examples the rate determining step was one of the sequential chemical reactions leading to a product. The rate-determining step can also be the transport of reactants to where they can interact and form the product. This case is referred to as diffusion control and, in general, occurs when the formation of product from the activated complex is very rapid and thus the provision of the supply of reactants is rate-determining.

Reaction Rate Constant

In chemical kinetics a reaction rate constant or reaction rate coefficient, k, quantifies the rate of a chemical reaction.

For a reaction between reactants A and B to form product C

$$aA + bB \rightarrow cC$$

the reaction rate is often found to have the form:

$$r = k(T)[\mathrm{A}]^m[\mathrm{B}]^n$$

Here $k(T)$ is the reaction rate constant that depends on temperature. [A] and [B] are the molar concentrations of substances A and B in moles per unit volume of solution, assuming the reaction is taking place throughout the volume of the solution. (For a reaction taking place at a boundary one would use instead moles of A or B per unit area).

The exponents m and n are called partial orders of reaction and are *not* generally equal to the stoichiometric coefficients a and b. Instead they depend on the reaction mechanism and can be determined experimentally.

Temperature Dependence

The Arrhenius equation gives the quantitative basis of the relationship between the activation energy and the reaction rate at which a reaction proceeds. The rate constant is then given by

$$k = Ae^{\frac{-E_a}{RT}}$$

and the reaction rate by

$$r = Ae^{\frac{-E_a}{RT}}[\mathrm{A}]^m[\mathrm{B}]^n,$$

where E_a is the activation energy, and R is the gas constant. Since at temperature T the molecules have energies according to a Boltzmann distribution, one can expect the proportion of collisions with energy greater than E_a to vary with $e^{-E_a/RT}$. A is the pre-exponential factor, or frequency factor (not to be confused here with the reactant A).

Units

The units of the rate constant depend on the global order of reaction: If concentration is measured in units of mol·L^{-1} (sometimes abbreviated as M), then

- For order $(m + n)$, the rate constant has units of mol$^{1-(m+n)}$·L$^{(m+n)-1}$·s^{-1}
- For order zero, the rate constant has units of mol·L^{-1}·s^{-1} (or M·s^{-1})
- For order one, the rate constant has units of s^{-1}
- For order two, the rate constant has units of L·mol^{-1}·s^{-1} (or M^{-1}·s^{-1})
- And for order three, the rate constant has units of L^{2}·mol^{-2}·s^{-1} (or M^{-2}·s^{-1})

Plasma and Gases

Calculation of rate constants of the processes of generation and relaxation of electronically and vibrationally excited particles are of significant importance. It is used, for example, in the computer simulation of processes in plasma chemistry or microelectronics. First-principle based models should be used for such calculation. It can be done with the help of computer simulation software.

Rate Constant Calculations

Rate constant can be calculated for elementary reactions by molecular dynamics simulations. One possible approach is to calculate the mean residence time of the molecule in the reactant state. Although this is feasible for small systems with short residence times, this approach is not widely applicable as reactions are often rare events on molecular scale. One simple approach to overcome this problem is Divided Saddle Theory. Such other methods as the Bennett Chandler procedure, and Milestoning are also developed for rate constant calculations.

Divided Saddle Theory

The theory is based on the assumption that the reaction can be described by a reaction coordinate, and that we can apply Boltzmann distribution at least in the reactant state. A new, especially reactive segment of the reactant, called the *saddle domain*, is introduced, and the rate constant is factored:

$$k = k_{\mathrm{SD}} \cdot \alpha_{\mathrm{RS}}^{\mathrm{SD}}$$

where $\alpha_{\mathrm{RS}}^{\mathrm{SD}}$ is the conversion factor between the reactant state and saddle domain, while k_{SD} is the rate constant from the saddle domain. The first can be simply calculated from the free energy surface, the latter is easily accessible from short molecular dynamics simulations.

Reaction Rate Theory

Already in 1940, Hendrik Kramers understood well the mechanism of climbing the reaction barrier in solution as governed by the thermal (Brownian) motions of the solvent. The above mentioned rare event problem did not play a role in the description he used, as the solvent effect was not described by explicit molecules, but rather by Gaussian random fluctuations $f(t)$ and by a linear friction force $-\gamma m \frac{dx}{dt}$ working on a reaction coordinate x with a mass m. A one-dimensional asymmetric double-well potential $U(x)$ represents the reactant and product well and the separating barrier along by the reaction coordinate, which enter with the solvent forces Newton's equation of motion of x in the form of a Langevin equation:

$$m\frac{d^2x}{dt^2} = -\frac{dU(dx)}{dx} - \gamma m\frac{dx}{dt} + f(t)$$

Here, γ is a constant friction rate, and the fluctuating force $f(t)$ denotes Gaussian white noise with zero mean, which obeys the fluctuation-dissipation relation,

$$\langle f(t) \rangle = 0,$$

$$\langle f(t)\ f(t') \rangle = 2m\gamma k_B T\delta(t - t')$$

The delta function in ensures instantaneous dissipation, which is known as the Markov property. From the time evolution of the probability density $p(x,v,t)$,, expressed in a so-called Fokker-Planck equation, he studied the reactant flux over the barrier for low and high friction γ (also referred to as damping, coupling or viscosity, and related to the diffusion coefficient via Einstein's relation: $\gamma = k_B T / D$). For moderate-to-strong friction he came to the cornerstone result in rate theory for the reaction rate constant:

$$k_R = \frac{1}{\omega_b}\left(-\frac{\gamma}{2} + \sqrt{\frac{\gamma^2}{4} + \omega_b^2}\right)\left\{\frac{\omega_R}{2\pi}\exp[-\beta E_b]\right\}$$

where R and ω_b the frequencies that confine the parabolic reactant well and barrier top respectively, as they are both of the form $U(x) = U(x') - \frac{1}{2}m\omega'(x - x')^2$.

Eyring Equation

The Eyring Equation, developed by Henry Eyring in 1935, is based on transition state theory and is used to describe the relationship between reaction rate and temperature. It is similar to the Arrhenius Equation, which also describes the temperature dependence of reaction rates. However, whereas Arrhenius Equation can be applied only to gas-phase kinetics, the Eyring Equation is useful in the study of gas, condensed, and mixed-phase reactions that have no relevance to the collision model.

The Eyring Equation gives a more accurate calculation of rate constants and provides insight into how a reaction progresses at the molecular level. The Equation is given below:

$$k = \frac{k_B T}{h} e^{-\frac{\Delta H^{\ddagger}}{RT}} e^{\frac{\Delta S^{\ddagger}}{R}}$$

Consider a bimolecular reaction:

$$A + B \rightarrow C$$

$$K = \frac{[C]}{[A][B]}$$

where K is the equilibrium constant. In the transition state model, the activated complex AB is formed:

$$A + B \rightleftharpoons AB^{\ddagger} \rightarrow C$$

$$K^{\ddagger} = \frac{[AB]^{\ddagger}}{[A][B]}$$

There is an energy barrier, called activation energy, in the reaction pathway. A certain amount of energy is required for the reaction to occur. The transition state, AB‡AB‡, is formed at maximum energy. This high-energy complex represents an unstable intermediate. Once the energy barrier is overcome, the reaction is able to proceed and product formation occurs.

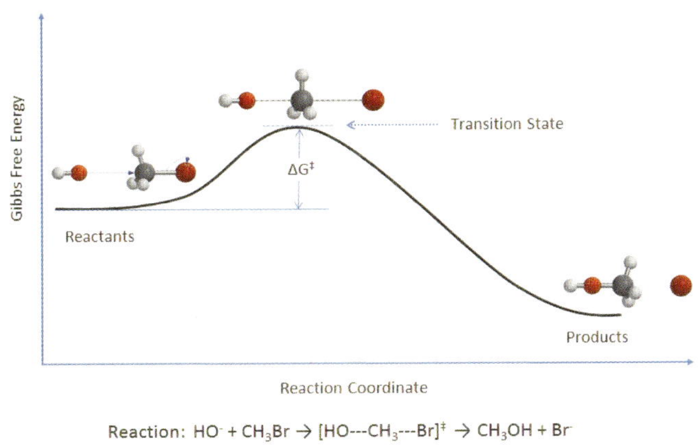

Reaction coordinate diagram for the bimolecular nucleophilic substitution (SN2SN2) reaction between bromomethane and the hydroxide anion

The rate of a reaction is equal to the number of activated complexes decomposing to form products. Hence, it is the concentration of the high-energy complex multiplied by the frequency of it surmounting the barrier.

$$rate = v[AB^{\ddagger}]$$
$$= v[A][B]K^{\ddagger}$$

The rate can be rewritten:

$$rate = k[A][B]$$

Combining the above two equations gives:

$$k[A][B] = v[A][B]K^{\ddagger}$$
$$k = vK^{\ddagger}$$

where,

- v is the frequency of vibration,
- k is the rate constant and
- $K^{\ddagger}$ is the thermodynamic equilibrium constant.

The frequency of vibration is given by:

$$v = \frac{k_B T}{h}$$

where,

- k_B is the Boltzmann's constant (1.381 x 10^{-23} J/K),
- T is the absolute temperature in Kelvin (K) and
- h is Planck's constant (6.626 x 10^{-34} Js).

Substituting above two equation:

$$k = \frac{k_B T}{h} K^{\ddagger}$$

Equation refref is often tagged with another term (M1−m)(M1−m) that makes the units equal with MM is the molarity and mm is the molecularly of the reaction.

$$k = \frac{k_B T}{h} K^{\ddagger} (M^{1-m})$$

The following thermodynamic equations further describe the equilibrium constant:

$$\Delta G^{\ddagger} = -RT\ ln\ K^{\ddagger}$$
$$\Delta G^{\ddagger} = \Delta H^{\ddagger} - T\Delta S^{\ddagger}$$

where $\Delta G^{\ddagger}$ is the Gibbs energy of activation, $\Delta H^{\ddagger}$ is the enthalpy of activation and $\Delta S^{\ddagger}$ is the entropy of activation. Combining above two equations to solve for ln $K^{\ddagger}$ gives:

$$ln\ K^{\ddagger} = -\frac{\Delta H^{\ddagger}}{RT} + \frac{\Delta S^{\ddagger}}{R}$$

The Eyring Equation is finally given by substituting Equation ($ln\ K^{\ddagger} = -\frac{\Delta H^{\ddagger}}{RT} + \frac{\Delta S^{\ddagger}}{R}$) into Equation ($k = \frac{k_B T}{h} K^{\ddagger}$):

$$k^{\ddagger} = \frac{k_B T}{h} e^{-\frac{\Delta H^{\ddagger}}{RT}} e^{\frac{\Delta S^{\ddagger}}{R}}$$

Application of the Eyring Equation

The linear form of the Eyring Equation is given below:

$$\ln\frac{k}{T} = \frac{-\Delta H^{\ddagger}}{R}\frac{1}{T} + \ln\frac{k_B}{h} + \frac{\Delta S^{\ddagger}}{R}$$

The values for $\Delta H^{\ddagger}$ and $\Delta S^{\ddagger}$ can be determined from kinetic data obtained from a ln $\frac{k}{T}$ vs. $\frac{1}{T}$ plot. The Equation is a straight line with negative slope, $\frac{-\Delta H^{\ddagger}}{R}$, and a y-intercept, $\frac{\Delta S^{\ddagger}}{R} + \ln\frac{k_B}{h}$.

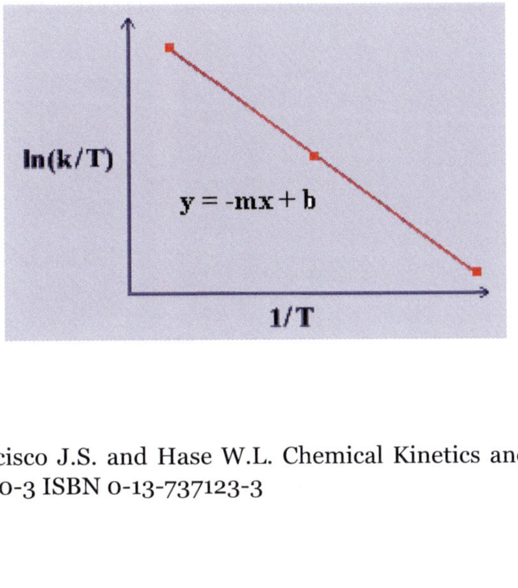

References

- Steinfeld J.I., Francisco J.S. and Hase W.L. Chemical Kinetics and Dynamics (2nd ed., Prentice-Hall 1999) p.140-3 ISBN 0-13-737123-3

- Dybkaer, R. (2001). "UNIT "KATAL" FOR CATALYTIC ACTIVITY (IUPAC Technical Report)". Pure Appl. Chem. 73 (6): 929. doi:10.1351/pac200173060927
- Chemical-kinetics, science: britannica.com, Retrieved 22 May 2018
- Ruthenium(VI)-Catalyzed Oxidation of Alcohols by Hexacyanoferrate(III): An Example of Mixed Order Mucientes, Antonio E,; de la Peña, María A. Journal of Chemical Education2006 83 1643
- Nelson, D. L. and Cox, M. M. (2000) Lehninger, Principles of Biochemistry 3rd Ed. Worth Publishing: New York. ISBN 1-57259-153-6
- Catalysis, science, classification-of-catalysts: britannica.com, Retrieved 9 July 2018
- Murdoch, Joseph R. (1981). "What is the rate-limiting step of a multistep reaction?". Journal of Chemical Education. 58 (1): 32–36. Bibcode:1981JChEd..58...32M. doi:10.1021/ed058p32
- Daru, János; Stirling, András (2014). "Divided Saddle Theory: A New Idea for Rate Constant Calculation". J. Chem. Theory Comput. 10 (3): 1121–1127. doi:10.1021/ct400970y
- Reaction-mechanism, science: britannica.com, Retrieved 29 June 2018
- Tinoco, K. Sauer and J. C. Wang Physical Chemistry. Principles and Applications in Biological Sciences. (3rd ed., Prentice-Hall 1995) pp. 328–9 ISBN 0-13-186545-5
- West, Anthony M.A.; Elber, Ron; Shalloway, David (2007). "Extending molecular dynamics time scales with milestoning: Example of complex kinetics in a solvated peptide". The Journal of Chemical Physics. 126 (14): 145104. Bibcode:2007JChPh.126n5104W. doi:10.1063/1.2716389. Retrieved 4 May 2018

Permissions

All chapters in this book are published with permission under the Creative Commons Attribution Share Alike License or equivalent. Every chapter published in this book has been scrutinized by our experts. Their significance has been extensively debated. The topics covered herein carry significant information for a comprehensive understanding. They may even be implemented as practical applications or may be referred to as a beginning point for further studies.

We would like to thank the editorial team for lending their expertise to make the book truly unique. They have played a crucial role in the development of this book. Without their invaluable contributions this book wouldn't have been possible. They have made vital efforts to compile up to date information on the varied aspects of this subject to make this book a valuable addition to the collection of many professionals and students.

This book was conceptualized with the vision of imparting up-to-date and integrated information in this field. To ensure the same, a matchless editorial board was set up. Every individual on the board went through rigorous rounds of assessment to prove their worth. After which they invested a large part of their time researching and compiling the most relevant data for our readers.

The editorial board has been involved in producing this book since its inception. They have spent rigorous hours researching and exploring the diverse topics which have resulted in the successful publishing of this book. They have passed on their knowledge of decades through this book. To expedite this challenging task, the publisher supported the team at every step. A small team of assistant editors was also appointed to further simplify the editing procedure and attain best results for the readers.

Apart from the editorial board, the designing team has also invested a significant amount of their time in understanding the subject and creating the most relevant covers. They scrutinized every image to scout for the most suitable representation of the subject and create an appropriate cover for the book.

The publishing team has been an ardent support to the editorial, designing and production team. Their endless efforts to recruit the best for this project, has resulted in the accomplishment of this book. They are a veteran in the field of academics and their pool of knowledge is as vast as their experience in printing. Their expertise and guidance has proved useful at every step. Their uncompromising quality standards have made this book an exceptional effort. Their encouragement from time to time has been an inspiration for everyone.

The publisher and the editorial board hope that this book will prove to be a valuable piece of knowledge for students, practitioners and scholars across the globe.

Index